# DRIVEWAYS, PATHS AND PATIOS

### A COMPLETE GUIDE TO DESIGN, MANAGEMENT AND CONSTRUCTION

# DRIVEWAYS, PATHS AND PATIOS

## A COMPLETE GUIDE TO DESIGN, MANAGEMENT AND CONSTRUCTION

### TONY McCORMACK

THE CROWOOD PRESS

First published in 2005 by
The Crowood Press Ltd
Ramsbury, Marlborough
Wiltshire SN8 2HR

**www.crowood.com**

**British Library Cataloguing-in-Publication Data**
A catalogue record for this book is available from the British Library.

ISBN 1 86126 778 9

**Dedication**
For my father, Tony senior, who first taught me how to lay flags and instilled in me a life-long love affair with the paving trade. He taught by example that integrity and hard work bring their own rewards.

**Disclaimer**
The author and publisher do not accept responsibility or liability in any manner whatsoever for any omission, nor any loss, damage, injury, or adverse outcome of any kind incurred as a result of the use of the information contained in this book, or reliance upon it. Readers are advised to seek professional advice relating to their particular property, project and circumstances before embarking on any building or related work.

FRONTISPIECE: Driveways have become significantly larger over time (Blockleys).

Unless otherwise credited, all of the illustrations in this book are by the author.

Designed and typeset by Focus Publishing, 11a St Botolph's Road, Sevenoaks, Kent

Printed and bound in Malaysia by Times Offset (M) Sdn Bhd.

# Contents

1    A Brief History of Paving ................................................................. 7

2    Spoiled for Choice ........................................................................... 16

3    The Design Process ......................................................................... 39

4    DIY? Contractor? Costs? ................................................................. 63

5    Costings ......................................................................................... 69

6    Tools .............................................................................................. 75

7    Pavement Layers ............................................................................ 80

8    Setting Out .................................................................................... 90

9    Drainage ...................................................................................... 116

10   Laying Block Paving ..................................................................... 127

11   Laying Techniques for Flags .......................................................... 153

12   Laying Other Materials ................................................................. 166

13   Steps, Ramps, Stepping Stones and Crazy Paving ........................ 175

14   Completion, Remedial Work and Disputes ................................... 180

Useful Links ........................................................................................... 186

Glossary ................................................................................................. 189

Index ..................................................................................................... 191

# CHAPTER 1

# A Brief History of Paving

## EARLIEST DAYS

Paving is one of the oldest of all the construction trades and has been used by man since long before we bothered to record history. At some unknown point in time, a human being, no less intelligent than ourselves, but living in very different circumstances, decided to improve the trackway he was using, perhaps throwing some dried rushes and twigs over a swampy stretch to help keep dry his poorly shod feet, and so formed the first 'improved' path. And maybe he scattered some sand and gravel, collected from a nearby beach or riverbank, on the area directly outside the family shelter, making it safer for children to play and cleaner for the clan to sit out and eat in the open air, and in the process unwittingly created the world's first patio. We shall never know, but it is a safe bet that others soon imitated the techniques, and word started going around that the bloke from the tribe over the hill was a bit flighty, and was using poor quality gravel or sub-standard rushes.

Pathways were essential to the development of human societies. They linked families and clans, tribes and kingdoms, providing trade routes and the main means of communication. Gathering areas – what we now often call patios – provided places for meeting and eating, locations where tales could be told, knowledge shared and gossip spread, so that much at least has hardly changed.

The earliest paths were 'improved trackways'. They

*OPPOSITE: Resealing products can rejuvenate a tired tarmac driveway and their application is an ideal DIY task. (Gardner-Gibson)*

had come into being naturally, as humans and other animals followed the easiest line through the landscape, skirting around the wettest patches, avoiding impassable obstacles, looking for the gentle gradients, and gradually developing from a stretch of trampled grasses to strips of bare earth or rock, slowly widened as travellers on two legs and four expanded the edges when their feet, paws and hooves sought drier ground. Dry matter would be added to improve the surface, sand, grit, gravel, rushes, leaves, and then someone would place a flattish stone or two, and we had the first flagger!

In the British Isles, some of the paths and trackways of pre-Roman times are still evident in the landscape. The Ridgeway of southern England along with the Icknield Way are, perhaps, the best known, following the higher, drier ground as they weave across the chalk landscape from Salisbury Plain to the Fenlands of East Anglia. Both are thought to have been used for at least 5,000 years and they probably go back much further than that.

Meanwhile, over in continental Europe the Romans took the principles developed in Mesopotamia, Egypt, Sumeria and Greece, developing and refining road-building technology to a new level. The Via Appia linking Rome with Capua is possibly the most famous of all Roman roads and has been dated to 312BC. The Legions had a well-developed strategy of having their cohorts build the roads as they progressed through a territory, continually expanding the Empire and the *Pax Romana*. Prisoners, slaves or general labourers followed in their wake, carrying out essential maintenance and repairs.

*Modern view of the Ridgeway near Uffington Castle, Oxfordshire.*

## *PAVIMENTUM* – THE ROMAN COLONIZATION

The Roman colonization of Britain brought a massive upgrade to some of the traditional trackways, but it was the construction of new routes, with their characteristic 'straight-line' alignment, that largely ignored topography, that has come to dominate what we now think of as Roman roads. These represented a quantum leap in pavement technology, comparable to the upgrading of a meandering country lane to a modern motorway. Not only were new construction techniques employed, but also a scientific, logical, methodical approach to paving was introduced. Carefully selected and graded materials, construction in definite layers, provision of drainage, and technically-competent setting-out by skilled workers known as *agrimensores*, all of which contributed to the basis of the modern civil engineering and surveying professions.

Here, for the first time in these islands, a national standard for construction was imposed, and so well-built were these roadways that several survive to this day, and many more established the route for some of our busiest modern highways. The key to the longevity of these pavements, other than military necessity, was the use of a cambered surface, with distinct layers of different materials forming the structure, and the provision of simple yet effective drainage at the edges. Added to these was the routing, which linked major towns and settlements, and so ensured that the roads were used, valued and therefore worth maintaining.

It is easy to be carried away by the progress such roadways represented and to overlook the fact that some of the techniques used would have been transferred to smaller scale projects. If the Fosse Way and Ermine Street were paved, then feeder or tributary roads would have adopted and adapted some of the principles involved, and access tracks to private dwellings and villas would similarly have borrowed some of the techniques. Indeed, it was in public buildings and the homes of the important citizenry that other forms of paving were used to great effect.

Busy town centres had paved footways and streets, with elementary separation for pedestrians and vehicles. These pavings were mostly of riven flagstones, blocks of dressed or pitched stone laid as setts, or loose gravelled surfacing, although there is a suggestion that timber may have been used in some situations – it's unlikely to have been stained blue with woad, though!

In the villas, hypocaust flooring used riven flagstone tiles or *tegulae* to form floors, and in the more important rooms the floors would be covered with mosaics, formed from thousands of tiny tesserae laid on a basic concrete known as *opus signinum*. Many of these mosaics were manufactured off-site, usually in southern Europe, prefabricated by highly skilled master mosaicists and selected from a catalogue by the client. They would be brought in to building projects when required and be laid by local tradesmen, much in the way that we do now with a feature patio.

Out of doors, function was far more important than form, and so courtyards were paved by using flagstone, setts and cobbles, with local materials dictating which style of paving was the most suitable. Where the local geology provided only poor or soft stone, beach or river cobbles, washed down from harder rock-beds miles upstream, could be used to provide a simple, hard-paved surface.

## 1,200 YEARS OF POTHOLES

The Roman control of Britain ended in the fifth century and with the loss of that control the last planned and methodical maintenance of the road network, so essential to ensuring military supremacy and the rule of law, also ended. The skills must have survived, but the political will was gone, and it would be a thousand years before any real political effort to maintain a serviceable road network would return.

The Dark Ages came and went, the Angles, Saxons, Jutes and others pushed their way in and paid scant attention to the finery and feats of their predecessors. They would use local materials to provide hard-standing for essential areas and fill in the odd pothole or two when it threatened to dislodge a wheel from their favourite cart, but paving, as a skilled trade, was relegated as the provision of food

*Although they seem coarse and uncomfortable to modern taste, pitched stone roads were the best option for hundreds of years.*

and shelter took precedence and the road network fell into decline.

The Vikings also came and continued in much the same manner. Their towns would use stone paving in the busiest sections, but it was the responsibility of individual landowners to provide any hard surfacing that was required. Merchants and craftsmen might put down a few setts, cobbles or flagstones outside their place of business, perhaps to minimize the amount of mud and worse carried into their premises on the shoes and boots of customers, and there would

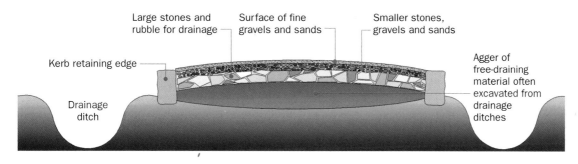

Large stones and rubble for drainage
Surface of fine gravels and sands
Smaller stones, gravels and sands
Kerb retaining edge
Agger of free-draining material often excavated from drainage ditches
Drainage ditch

*Actual construction methods varied throughout the Roman Empire to make best use of local materials. This cross-section illustrates a typical construction for a British road which would be surfaced with gravels, rather than paved with flagstones or setts.*

be flagstones on the floors of the more upmarket properties, but most homes would rely on hard-packed, earthen floors with straw changed at irregular intervals.

The Normans brought back the traditional skills of masonry, driven by their intensive programme of castle building, and the craft of the stone workers was employed for both vertical structures and for floors and courtyards. The basic tools used by Norman masons and paviors remain with us still. Hammers and wedges to split the stone, chisels and punches to dress it, and mallets to settle the stones into the bedding.

And so it went on. It was during the Tudor period, in 1555, that an Act of Parliament created the position of Surveyor of Highways, which was unpaid, unpopular and ineffective. Parishes were required to maintain their area's roads and a bursary was issued to the local population who were required to 'mend their ways'.

## TURNPIKES, TAR AND TECHNICAL IMPROVEMENTS

It took until the late seventeenth century before properly maintained turnpike roads were established. The tolls imposed were used for their upkeep and extension, and the first successful road of this type was the Great North Road, now more commonly known as the A1. Other turnpike roads were developed throughout the later seventeenth century and their financial success led to their being expanded significantly throughout the following century, a period that saw ever improved designs and construction methods. However, as the road network expanded, so did the traffic, and contemporary construction methods struggled to cope. Surfaces became rutted and potholed; some of the turnpike trusts were less than honest when apportioning funds, and for the first time in centuries the quality of roads deteriorated.

Luckily, a number of men applied their thoughts to these matters and gradually, thanks to their efforts, significant improvements were made, improvements that formed the basis of modern road and pavement construction technology.

Despite his inability to see what he was doing, Blind Jack of Knaresborough, more properly known as John Metcalfe, created 250km (160 miles) of turnpike road in Yorkshire that relied on a cambered surface and drainage ditches at the edges. Sound familiar? The master civil engineer Thomas Telford understood the critical need to drain pavements properly, and so developed a system using pitched stones overlain by layers of progressively smaller sized crushed stone and gravels, with camber or crossfall to drain surface water to roadside ditches.

John MacAdam developed his now legendary 'macadamized' system while employed as surveyor to the Bristol Turnpike Trust in 1816. This relied on using a fine crushed stone to bind together granite chippings and/or gravels laid over larger crushed stone, with a pronounced camber to the structure that would ensure water was shed to either side of the pavement, where it could not compromise the foundations. His fame spread: he was appointed Surveyor-General for Metropolitan roads and, before long, his methods were adopted throughout the western world.

By the time Victoria came to the throne, there was an extensive network of good quality roads, and it was the vehicles in use that limited speed and journey times rather than the surfacing. With a lightweight carriage and a good team of horses, speeds of 12mph were typical, and over a period of about fifty years, the journey from 'Cottonopolis' (Manchester) to London was cut from four days to a little over 36hr.

By this time, towns and cities had well-developed paving and surfacing for both carriageways and footways. Flagstones were hauled from the Pennine quarries all the way to London. Granite from Cornwall, Aberdeen and Wicklow was popular throughout the islands, and those clever Victorians were experimenting with new materials. Complaints regarding the noise from horses' hooves and iron-rimmed wheels as they traversed the granite setts of the streets of Westminster and London led to experiments with rectangular and hexagonal wooden 'setts', and then with the new wonder material of the age, rubber. Timber had a relatively short lifespan, three to four years was typical, but it was popular because traffic noise was significantly reduced, so much so that properties on wooden-paved streets attracted premium rents and lodging rates.

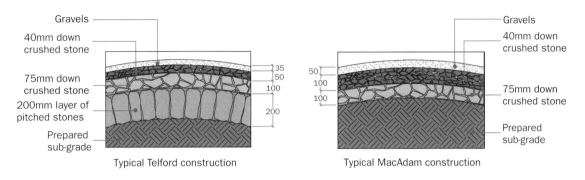

Labels (left diagram): Gravels; 40mm down crushed stone; 75mm down crushed stone; 200mm layer of pitched stones; Prepared sub-grade. Dimensions: 35, 50, 100, 200.

Typical Telford construction

Labels (right diagram): Gravels; 40mm down crushed stone; 75mm down crushed stone; Prepared sub-grade. Dimensions: 50, 100, 100.

Typical MacAdam construction

*Telford's construction, on the left, was refined and enhanced by MacAdam; their work remains at the heart of modern pavement construction technology.*

## NEW MATERIALS

Tar macadam has several alleged 'inventors'. Its immediate predecessor, asphalt, is documented as being used in Paris in the 1850s and seems to have appeared in London and Dublin within a decade. Tarmac, as we think of it today, was first patented in 1902 by Edgar Purnell Hooley, a surveyor from Nottingham. The tale is that he had noticed a wonderfully smooth section of roadway near Denby in Derbyshire, and found that, where a barrel of tar had been accidentally spilled, it had been covered with furnace slag. His patent, which became the foundation of the modern company Tarmac plc, covered the heating of the tar and the subsequent addition of ground slag, macadam and small stones.

The other great nineteenth century advance in paving materials came with the development of Portland cement. In 1824, Joseph Aspdin, a Yorkshire stonemason and bricklayer, was granted a patent for the cement he had developed from a mixture of crushed limestone and clays that was heated in a kiln and then ground to a powder. The product hardened when water was added, and he at first thought it resembled the popular building material Portland stone, hence the name.

Lime and gypsum mortars had been known since the time of the pharaohs, and the Romans had developed pozzolanic concrete by using the volcanic ash found near Pozzuoli in Italy. In 1756, John Smeaton had developed what is considered to be the first modern concrete by mixing pebbles with powdered brick dust. Portland cement took this to new levels of possibility, as its tremendous compressive strength and versatility redefined construction technology.

Portland cement concrete became a bedding material for many other pavings and surfacings, and experiments with casting led to the development of regular, uniform concrete flags as replacements for the somewhat haphazard stone flags that often had to be hauled hundreds of miles to where they were needed. The first 'standard' for concrete flags was published in 1929, and from thereon, there was a steady decline in the use of stone flags.

The advent of the railways had greatly improved logistics, and standardized building materials were becoming more widely available. Good quality clay bricks for masonry and paving – which had previously been restricted to their own local area – could now be delivered anywhere on the rail network; their use spread, along with that of the decorative clay edging tiles and diamond-pattern pavers that are now highly prized in reclamation yards.

Concrete block paving came to us from Europe. Devastated by two World Wars, the need to rebuild towns and cities quickly and cheaply led to the development of paving blocks that could be manufactured by pressing concrete instead of kiln-firing clay, and could be laid faster and more easily than concrete flagstones. Block paving first appeared in Britain during the 1970s but really took off in the 1980s, when local authorities looked to pedestrianization as a way of rejuvenating tired but busy towns and cities.

*Concrete reproductions of items such as reclaimed railways sleepers have become popular as casting technology has improved. (Longborough Concrete)*

*Driveways are an essential part of a home and have become significantly larger over time. (Blockleys)*

At that time, tarmac was popular but was seen as soulless, while concrete flags had become the norm, used anywhere and everywhere for public footpaths, driveways, town centres, civic squares and shopping centres. Block paving offered flexibility, modern styling and, most of all, colour. Town after town, city after city ripped up the old paving and installed block paving in the rush to revitalize their municipalities and attract shoppers and commerce.

What householders saw in the towns, they liked, and wanted it on their driveways. The 1990s saw concrete block paving become the most popular form of paving for new and reconstructed driveways. As personal wealth and leisure time increased, those same householders began to look at ways of improving their gardens as well as their driveways, and the great patio culture began, driven by aspirational television makeover and lifestyle programmes. The range of wet-cast products grew each year to meet a seemingly insatiable demand for new products, which were increasingly new versions of old products: wet-cast concrete copies of old stone flags.

As the millennium ended and a new century began, the breadth of paving materials available had never been wider, but still the driveway- and patio-buying public looked for more. Stone flags from India, granite setts from China, new textures and formats for block paving, concrete castings of old timber sleepers, patterned and stencilled concrete … it is an exciting time for paving.

## MODERN TASTES

The use of residential paving has changed over recent years. Originally it kept our feet clean between the public footpath and the house. A simple path of tiles or flagstones was sufficient. The growth in car ownership over the last fifty years has meant that a driveway has become an essential component of a home, and now a single driveway is not enough: we need double or triple driveways since many homes have more than one car. At the back of the house, the garden is now a lifestyle statement and an essential part of that styling is a patio. Socializing in the garden is now an accepted part of our lives and we spend more and more of our income on 'leisure' with each passing year.

The presence of an acceptable driveway and a reasonable patio are key selling points when a property is put on to the market. A driveway has to have

*OPPOSITE: Terracotta pavings bring a touch of the Mediterranean to our gardens. (Marshalls)*

'kerbside appeal'. The ability to park a car next to the house is not enough; it has to portray a lifestyle, it has to make us want it, to envy the generous width, to desire the security it offers, to long for the way it leads the eye to the well-appointed and impressive main entrance. Two strips of concrete flags with gravel chippings between them is no longer adequate. A driveway tells the world about the status of its owners, their hopes, their expectations, their standards and their way of life. An attractive and functional driveway can make the difference between a house selling in days and its languishing on the market for weeks. A shabby, weed-infested, miniature rectangle of dirty tarmac, crumbling concrete or broken flags hints at what lies behind the front door, at what sort of people live there. The driveway is the entrée, the starter course, and as such it whets the appetite and sets the tone for the remainder of the meal that is a home.

The patio is a more private area and, unlike the driveway, is normally seen only by invitation. To be invited on to the patio is to be admitted into the inner sanctum – you are an honoured guest of the family, a friend, a confidante, you are accepted. The milk and the bins, the post and the papers, all come to the house via the driveway, but only the favoured few make it on to the patio, it has become that most awful of television makeover clichés: a room outdoors.

But it is not a room, it is a space, and a transitional area between house and garden; it performs some of the socializing functions of the house, but it has to blend with the garden. It needs to suit the architectural styling of the house, yet continue the themes of the internal décor. It is the crossroads between indoor and outdoor, between form and function, between duty and leisure. It needs to provide a viewpoint for the garden, a taster of what to expect, a summary of its style or theme, but it also needs to be easily maintained and to make the best use of the space available.

Fashion has come to the patio. The types of material used will help to define it as classic, retro, modern or avant-garde, and dictate what type of planting, what type of furniture, what type of lighting should be associated with it. As will be seen later, choosing materials and designing a patio require an acknowledgement of the existing, an understanding of the possible, and an aspiration for something that defines the home and the family. A classic styling will probably never be out of fashion, but may be seen as too safe or too conservative. Retro places the patio in a specific period and imposes a limitation on what can be used and what can be planted, but evokes a particular sense of period. Modern styling is clean, simple and uncluttered, but for how long will it be modern? 'Decking' was modern for a few months in 1999 and

*A patio has to perform many functions. (Bradstone)*

is now well past it, according to some. Avant-garde forces people to think, to ponder, to provoke a reaction, to commit them to liking or disliking – there is no middle ground; it is love or hate, and is so highly individualistic that it is possibly not the best look for a house that is being put on the market.

What is your style? Do you know? Does it matter?

## FUTURE TRENDS

Fashion is fickle, it seems to emerge fully formed from the ether. Is it pushed forward by the manufacturers or are they dragged along in its wake? Do the media pundits and makeover programmes show us what we want, or what they want us to want? Predicting what will be fashionable next season is a gamble, even for those of us immersed in the paving trade, but there are some general trends that it may be assumed will be with us for the next decade or so.

Natural materials are in: flags of sandstone, quartzite, porphyry and limestone, setts of granite, whin and diorite, terracotta and travertine tiles, tumbled clay pavers, self-binding gravels. New materials are favoured, but there is a steady level of demand for the reclaimed and a growing demand for the recycled. The styling is modern, with touches of tradition, but individualistic and quirky. Good looks, plenty of space, high quality workmanship and ease of maintenance are not enough, a modern patio has to be a conversation piece.

Concrete products are fighting a rearguard action and are being driven in one of three directions: to extremely high quality and faithful reproduction of natural materials; ultra-modern chic, clean lines and minimalist styling; or dirt-cheap, budget bargains. There is a need for all three and none will ever dominate. There needs to be such offerings to provide the range of choice we now demand, even if it is only to compare and contrast.

Tarmac will still have its adherents because it is simple and functional; but new, updated looks will be sought – coloured macadams or resin-bound surfaces to inject a more playful or idiosyncratic styling. Decorative concrete will similarly remain moderately popular, but the standard of workmanship is the biggest problem facing that sector of the industry, and, unless it improves its image and stops shooting

*Clean lines and classic colours create a contemporary look for this patio. (Stonemarket)*

itself in the foot because of the antics of a few installers, it will never be more than a minority taste.

Overall, the public are looking for quality rather than a bargain. There will always be a cost-obsessed core looking for 'cheap', but, increasingly, the public cares more about a high standard of workmanship than a rock-bottom price. The past fifteen to twenty years have taught them to be wary and sceptical of contractors, and the trade has only itself to blame. We have abandoned traditional apprenticeships and forgotten how to train the flaggers, kerbers and block layers of tomorrow. We in the trade allow anyone to set themselves up as a paving contractor and then complain about the number of cowboys in the trade. We believe that we work to high standards and act with integrity, but we cannot organize an effective trade body to ensure that we actually meet and maintain those laudable ideals.

The rest of this book looks at how things should be done, whether by contractor or the DIY-er. It aims to promote best practice and to provoke discussion on design and materials and methodologies. It can never be fully comprehensive, but I hope that it inspires, encourages and helps in some small way to ensure that the reader gets the patio, path or driveway that he or she desires and deserves.

# CHAPTER 2

# Spoiled for Choice

## INTRODUCTION

Almost fifty years ago, when my father started his paving company, there were flags, there was concrete and there was tarmac. Choice was limited, but simple, straightforward and uncomplicated. Now, in the early part of a new century, we have dozens of manufacturers, each producing glossy, full-colour brochures with page after page of choices, options, variations, colourways, sizes, stylings and possibilities. We have never had it so good, as someone once said, but so much choice brings problems in itself. Just which do you choose, what is best for you and your project, and who are the better manufacturers? This section aims to explore the more popular products, show what is available, give tips and hints on how to select the most appropriate, and give a general push in the right direction.

## BLOCK PAVING

Concrete block paving of one form or another is the most popular choice for residential paving and has been so since the early 1990s, when economies of scale resulting from the considerable investment in new production plants by a handful of major manufacturers brought the price per square metre down to a more realistic level. Demand grew and so did the number of 'block paving specialists' that suddenly appeared on the scene.

At the time of writing (early 2005), block paving sales for the United Kingdom are expected to total around 28 million square metres ($m^2$), and it is domestic driveways, paths and patios that comprise

the bulk of this figure, approximately 17 to 18 million $m^2$ per year.

From the basic, original, rectangular 'brick' format, the range of options has expanded and the property owner is now faced with a sometimes bewildering array of choices – of colours, of sizes, of shapes, of textures, of thicknesses, of styles … so much so that some manufacturers are now limiting their production to just the most popular pavers in an attempt to reduce confusion. However, not a year goes past without some new paver entering the market, and what follows is a brief examination of the main types.

## Concrete and Clay

The biggest distinction between the types of block and brick pavers is between those manufactured by pressing concrete into a mould, what we call concrete block pavers or CBPs, and those pavers created by firing clay bricks in a kiln. They are compared in the table overleaf.

### Concrete Pavers

**Methods of manufacture** Britain differs from most other countries in that the majority of the pavers turned out by manufacturers are a 'through colour block': they are made from one mass of concrete with colouring right through the block. Pavers from Ireland, continental Europe and elsewhere are usually 'face mix blocks'; these use a high quality concrete with all the necessary colouring dyes in a separate 8–12mm layer on the top surface of the blocks, with an economy 'backing' used for the lower layer that will never be seen once the paving is complete.

*A small selection of the many types of block paver now available. (Tobermore)*

There is no compromise of quality with Face Mix blocks. They are not prone to frost damage nor delamination. They have been used without problems in Europe for forty years or more, and our continental neighbours endure far harder frosts than we ever get in our mild, maritime climate. The face mix is pressed on to the backing mix within seconds: there is no 'joint' between the two, no plane of weakness. Both Face Mix and Through Colour blocks

*Face mix has all the colour concentrated in one layer on the top 'face' while through colour blocks have the colour dispersed right through.*

Comparison of Concrete and Clay Pavers

| | Concrete block pavers | Clay pavers |
|---|---|---|
| **Cost** | Low, from £7/m² | Normally more expensive, £12/m² and upwards |
| **Pros** | Large choice of shapes, colours, textures and styles. Dimensionally accurate units for accurate laying. Range of block depths to suit all applications. | Exceptionally hardwearing. Natural non-fade colours. Some patterned formats available. |
| **Cons** | Colours prone to fading. Aggregate can become exposed with wear and weathering. | Sizes variable due to firing process. Prone to colonization by mosses. Mostly square or rectangular. Relatively hard to cut. Limited choice of block thickness. |
| **Lifespan** | At least 20 years | Clay pavers from the sixteenth century are still in use. |

*The red-multi 'brindle' colour is the favourite choice for most residential driveways; it is usually edged with blocks of charcoal, buff or both. (Formpave)*

have to meet the same standards of strength and durability. In fact, there is a good argument to be made that Face Mix blocks are a superior product: the expensive dyes can be concentrated in the upper layer, where they are needed, and not dispersed through the whole block. Similarly, the face mix can be manufactured from the highest quality aggregates, while the backing mix can include reclaimed or re-cycled aggregates. As long as the strength require-ments are met, there is no structural problem in using a lower quality concrete for the backing mix. One British manufacturer estimates that face mix products offer a cost saving of around 70p per m², compared with through colour blocks using high-quality aggregates and dyes throughout.

**Standard pavers** The most popular format for the concrete paver in Britain and Ireland is a 200 × 100mm rectangular block, having a depth of 40–80mm, with 50mm being the most popular for residential driveways. The blocks are available at a relatively low cost (£7–10 per m²) in a range of single colours and multi-colours. The most popular multi-colour is a red-grey mix commonly named 'brindle' and this is often combined with edge courses of a single colour, such as charcoal or buff, to create an attractive contrast.

For the main paved area of a driveway, termed the 'body' or 'field' of a pavement, a multi-colour block is a good choice. The mottled colouring is better able to disguise minor stains and markings, whereas single colours often emphasize any spot or mark that differs in any way from the rest of the paving. However, the manufacturers supplying the British market have two

different ways of creating the mottled, multi-colour effect. Most incorporate two or more coloured concretes within each block, and so no two blocks are the same; each contains a variable proportion of the coloured concretes in an infinite variation of pattern within each block. A minority of manufacturers achieve their so-called multi-coloured effect by mixing blocks of slightly different tones. So, for a brindle multi-colour, the manufacture may provide five or six different shades of red and grey blocks that need to be randomized before being laid to avoid a patchy appearance within the finished pavement.

Standard blocks are most commonly laid in one of three patterns: herringbone, stretcher (running) bond, or basketweave. Of these, herringbone offers the highest degree of interlock and is therefore the 'strongest' pattern and the one most commonly recommended for driveway use; stretcher bond offers less interlock, as there is potential for movement between the courses; and basketweave is the weakest of the three, as there is potential for pattern creep in both the transverse (side-to-side) and the longitu-dinal (top-to-bottom) direction. Other patterns are possible, but they are more complex to lay and often weaker than the basic herringbone pattern.

Standard blocks offer a 2:1 format – each block is twice as long as it is wide. There are variations based on 3:1 and even 4:1 formats, and, while these are fine for driveway and patio use, they may not meet the requirements of the relevant British Standard which requires a length-to-width ratio of not more than 2:1. However, some people feel that these alternative formats offer enhanced aesthetic appeal.

**Tumbled pavers** Some years ago, in a faraway land, a man sat on a beach and watched the waves tumbling a brick back and forward, rounding off the corners, distressing the edges, imparting an aged, timeworn character to even the newest of bricks. In a flash of inspiration he realized that the tumbling action created by the waves could be replicated by machinery and used to age and distress concrete blocks, particularly paving blocks; and so the tumbled block came into being. The most famous of these products traces its roots back to the original inventor and is licensed to manufacturers all over the world, who manufacture the blocks in sizes and

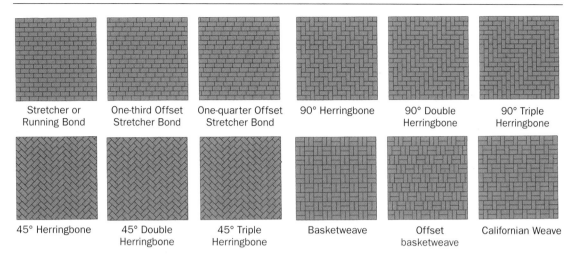

| Stretcher or Running Bond | One-third Offset Stretcher Bond | One-quarter Offset Stretcher Bond | 90° Herringbone | 90° Double Herringbone | 90° Triple Herringbone |

| 45° Herringbone | 45° Double Herringbone | 45° Triple Herringbone | Basketweave | Offset basketweave | Californian Weave |

*A selection of popular laying patterns for 2:1 ratio rectangular blocks.*

*ABOVE: Tumbled blocks laid to a sweeping arc. (Tobermore)*

*RIGHT: Tumbled blocks are often used to create circles, either as features 'inserted' into an area of coursework or, as in this image, as the central feature of a paved area. (Brett)*

colours to suit their local market and then sell them under the registered trade name *Tegula*, from the Latin for a tile.

The popularity of the *Tegula* block inspired many imitators, that are sometimes referred to as 'rumbled', but the basic premise remains that a block is rolled over and over in a rotating, cylindrical drum, bashing against the drum and against the other blocks, until it emerges some minutes later in a thoroughly bedraggled, but value-added condition. Since the original development of the *Tegula* system, alternative methods of distressing blocks have been developed, and some have made their originators quite wealthy, but the end result is generally the same – a bashed and battered concrete block with its corners and arrises (edges) missing.

There are some exorbitant claims made for these tumbled pavers. Manufacturers grandly claim that they 'replicate time-worn stone pavings from yesteryear', which is somewhat fanciful, but there is no denying that the finished product certainly does have its charms and is much more sympathetic to older properties, or 'homes of character', than the modern, geometric styling of standard blocks. It also avoids the often pathetic attempts to replicate traditional setts. The big selling point for these tumbled pavers is that they offer all the advantages of modern block paving, such as cost savings and ease of laying, but may be used with some success on properties where stone setts or flagstones would previously have been the only choice.

**Decorative pavers** This category is where we lump all the blocks that do not fall into one or other of the previous groupings. It includes those that are 'different' in some way, which generally means in texture, shape and size, or a combination of these. Non-rectangular shapes are generally manufactured in moulds, just as are the standard and the tumbled pavers, and there is usually an underlying mathematical principle that allows them to be mixed and matched with blocks from the same range into suitable patterns, or with blocks from complementary ranges to create distinctive layouts.

Special textures are most commonly formed by 'secondary processing', which involves subjecting the block to additional treatment following its initial manufacture. It may be washed or polished to expose a decorative aggregate, or shot-blasted or bush-hammered to create a more stone-like surface. Secondary processing, which includes the tumbling process mentioned above, always involves extra manufacturing costs, and these are reflected in the retail price of the pavers.

*Clay Pavers*

**Methods of manufacture** Clay bricks have been used as paving materials for hundreds of years. Bricks from the fifteenth century can be seen at Hampton Court to this day, which bears testament to just how hard-wearing fired clay can be when used as a paver. Note how the trade often refers to clay 'bricks' rather than 'blocks', although they are also known as

*Some blocks mimic the slightly domed surface often found with gritstone setts. (Stoneflair)*

*A 4:1 ratio, textured block paver laid in stretcher bond. (Tobermore)*

*Clay pavers come in a wide selection of colours. (Baggeridge)*

pavers, paviors, pavoirs, pamments, or simply as 'clays'.

Originally, clay pavers were made in the same way as other bricks, what we now call facing or common bricks. Experience led to certain types of clay and methods of firing being identified as resulting in a type of brick that was durable when used 'on the flat' and in permanent contact with the ground. It is worth noting that not all clay bricks are suitable for use as pavers. Many new and reclaimed facing (house) or common bricks do not have an adequate level of frost resistance to cope with their being used in damp conditions, which often results in the 'spalling' of the surface. The worst of them degrade to

a mush after a few years, which is why the use of facings or 'house' bricks is rarely recommended for use in paving projects. Further, many facing or house bricks do not have a true 2:1 ratio that enables them to be laid in accurate interlocking patterns.

Most modern clay pavers are manufactured in production plants dedicated to manufacturing 'facings'. The methods and materials used are similar, and so their manufacture complements the production of 'house' bricks and can be a profitable line for factories with access to suitable clays. There are only a handful of British factories producing clay pavers, and only one or two have facilities exclusively dedicated to the production of pavers.

The level of demand is barely a tenth of that for their concrete cousins, and this is largely due to the historical price difference. However, as CBP manufacturers have brought out high-value products (such as the tumbled and decorative blocks mentioned earlier) clays now find themselves as a mid-price product and this has helped to stimulate demand.

Different colours of clay pavers are manufactured by blending different clays before firing. The most common method of manufacture involves the prepared clay being extruded and then sliced into brick-sized pieces. This slicing of the extruded clay is responsible for the classic 'dragwire' texture observed with many modern pavers. A small proportion of pavers, known as 'stock pavers', are moulded – the clay is literally thrown into a mould of the required size and shape to create a highly distinctive look and texture.

The raw bricks are arranged into stacks and allowed to dry out before being transferred on a 'kiln car' to a gas-fired tunnel kiln where they are fired at high temperatures (1150°C) for an extended period. The position of an individual brick within a stack, and the position of that stack on the kiln car, the firing conditions of the kiln (known as oxidation or reduction firing), and the 'top firing' temperature, all have a direct affect on the finished appearance and colouring of the brick.

Once fired, the bricks are allowed to cool in a controlled manner while still inside the kiln. Once cooled, however, they can be packaged, shipped and used immediately. All this care and time, not to mention the fuel to fire the kiln, make clay pavers more expensive than concrete blocks, but the cost is often of secondary importance on projects where clays are used. It is their durability and colour-fastness that makes them attractive to designers and specifiers. After the hit-and-miss success with concrete pavers on the pedestrianization schemes of the 1980s, clay pavers offer a surface that will be the same colour twenty years on as it is today, and will not be worn away by stiletto heels and delivery vehicles to reveal a less-attractive and differently-coloured aggregate.

With a smaller market share, there is less demand for variety from clays than is the case with concrete blocks, but, again, the pavers may be arbitrarily divided into the three groups discussed above: standard, tumbled, and decorative.

In terms of colour, there are four main groupings, but an almost continuous spectrum of colour from buffs to brown, through red and beyond to the deepest indigo blues is available. As the clay used in their manufacture is a natural material, it is almost infinitely variable, which is part of its charm, and this variability enables local or regional colouring to be used as part of a larger hardscape.

Reds are probably the most popular colours, containing true 'brick' reds and a huge range of blends that incorporates russets and browns, tans, oranges, beiges and even blacks. The buffs and browns offer softer, muted, more organic and less strident tones and hues. Buffs bring lightness to darker or shady areas, while browns are earthy and sympathetic yet offer an ability to disguise minor staining. The blues typically hail from Staffordshire, where the local marl clay produces an incredibly strong and heavy brick. Many of the nation's railway bridges and tunnels were constructed from bricks manufactured from the Staffordshire blue. From a designer's viewpoint, the blues are usually considered too dark to be used in creating large areas of colour, and so their most common use is as a contrast to other colours, as band or edge courses.

**Standard clays** These are the basic, rectangular 'bricks'. They may be chamfered or chamferless; they may have spacer nibs or be nibless; their dimensions may be in the ratios 2:1, 3:1, 3:2 or some other value, but they are typically bricklike.

**Tumbled clays** Tumbling or distressing is a relatively modern development, taken in response to the overwhelming success of tumbled concrete pavers. The tumbling of clay pavers imparts a genuine aged appearance. When concrete blocks are distressed, the process exposes the internal concrete structure, which is not always as attractive as the untrammelled surface. However, with clays, all that is revealed is more fired clay, exactly the same as that on the surface, on the sides and on the base.

Tumbled clays have an uncanny ability to look reclaimed from the moment they leave the factory. They do not suit every job, but for projects in

*Tumbled clays bring an added texture to a pavement. (Marshalls)*

*Diamond-pattern clay pavers have been a favourite for over a hundred years. (Baggeridge)*

gardens or where they are part of a larger softscape, tumbled pavers bring a natural and gentle feel to what could otherwise be a hard and austere surface.

**Decorative clays** There have been decorative clays for decades, longer than we have been using concrete pavers of even the most basic form. One of the most popular forms has been the 'diamond' paver, a thinner paver that is right on the border between tiles and bricks (which we in the trade arbitrarily set at 30mm). There are also the 'stable' pavers, sometimes known as 'chocolate block' pavers because of their distinctive form. Their natural durability and resistance to wear made them an ideal choice for stable yards that experience severe abrasion from hooves. Previously, such yards had been cobbled with hard river or beach stones, or paved with hardstone setts, both of which are labour-intensive and difficult to lay (explained below). Mass production and improved logistics thanks to the canals and railways made these tiles widely available and much easier to install.

The idea of using a pattern or moulded texture never went away, but over recent years new designs have been produced, some more successful than others, but the old favourites persist.

## FLAGS AND SLABS

They are called flags in northern Britain, shortened from the older term 'flagstones' which is used to describe the large, flattish sections of rock that have been used for centuries to provide a serviceable surface for man and beast. In the south of England, and a few other parts of these islands, they are referred to as 'slabs', derived from the term used to describe large bays of concrete. Whatever they are called, they have an important role in paving of all scales, from garden paths and patios to the largest urban hardscape scheme. There are two classes of flag: natural stone and concrete. The types of stone or concrete may be used to further categorize the flags, but for now, classifying them as stone or concrete will suffice.

## Stone Flags

Stone flags vary according to the local geology, and some types of rock are so suitable for use as flagstones that they have been exported beyond their regional boundaries to all corners of the country, and, indeed, to all parts of the world. Yorkstone from the Pennines of northern England can be found on projects in Gibraltar, Singapore, Australia and North America. We now see sandstones and limestones from India being imported to Europe, but this is less to do with the quality of the stone and more to do with the cost of labour in developing nations.

Stone flags do not come out of a quarry as perfect rectangles of regular thickness. They have to be worked, either by means of tough machinery or time-consuming, manual labour. Once, most flags were handworked, what is known as 'quarry fettled', but modern technology crept in and the vast majority of

*A selection of colours and textures readily available from local stockists. (Stonemarket)*

British and Irish flagstones are now cut and finished by machine. In contrast, much of the flagstone from Asia and Africa is still quarry fettled and the natural look has helped to establish it in the modern market-place for garden and patio paving.

## Concrete Flags

It was the difficulty in working stone, both in terms of machinery and manpower, combined with its inherent variation in character and the logistics of moving it from where it was quarried to where it was needed, that originally drove the market for a more amenable alternative. With the production of reliable cements from the 1900s onward and consequential developments in concrete technology, it was not long

*Patio area paved with a grey-toned imported sandstone.*

before someone started knocking out a concrete version of a basic flagstone to meet demand. Suddenly, there were strong and durable flags that were of regular size and regular thickness, of a consistent quality, and manufactured locally at a fraction of the cost involved in quarrying and hauling stone across the country.

The technology was adapted and adopted and before long we had two distinct technologies: pressed concrete and cast concrete. The pressed concrete flags tended to be somewhat plain, but they were tough and durable, while wet-casting allowed for more decorative forms. Although both forms rely on placing concrete into a mould, the pressed flags, as their name suggests, rely on the barely moist concrete being pressed and squeezed into a steel mould, thereby thoroughly compacting the concrete and forcing out any surplus water. The result is a flag that is perfectly flat and accurate on all six faces. The casting technique uses a wetter concrete, in a flexible plastic or polyurethane mould, that is rigorously vibrated to shake out any air bubbles and ensure the maximum contact between concrete and mould. The face that will become the top, plus the four edges, all take the impression of the mould, but the base, which is the top face during moulding, is left open and finds its own level and texture.

Pressed flags became the norm for local authorities to use in the construction of public footpaths and town centre shopping areas. The sizes were based on old imperial measures and so there were 'two-by-twos' and 'three-by-twos', and were 2 or 2½in thick. It took a fit man to lay them, but they were eminently more manageable than their stone predecessors. Coercing 60+kg (125lb) of concrete to lie flat took muscles, and many flaggers suffered with back and knee problems by the time they reached middle age. Meanwhile, the considerably smaller and more manageable decorative, wet-cast products were manufactured to serve the home improvement market and had to be within the handling capabilities of the average home handyman.

Now, in the twenty-first century, health and safety requirements have caught up with the plight of the poor flaggers, and the size of pressed concrete flags has been reduced. The old '3 × 2' became 900 × 600mm in the metrification of the 1970s, and now 450 × 450mm is the largest hand-laid format.

## Types of Flag

*Pressed Concrete*
The bulk of pressed flag production is directed at the 'specification' market, that part of the industry dealing with public and commercial works. These flags are plain and functional, which is all that is wanted from them – utilitarian, rather than decorative. However, there are some decorative flags, targeted at both the specification and the home-improvement markets. The simplest decorative forms are those pressed into a mould with a bas-relief upper face. This technique is used to impart a basic riven-effect surface, but the need to use a relatively dry concrete in press manufacturing techniques means that the degree of realism is limited and the flags produced tend to be aimed towards the budget end of the market.

More decorative effects are possible with pressed flags by applying 'secondary processes'. These take a standard pressed flag and treat the surface in some way to impart a new texture. The two most popular techniques involve the roughening of the surface by means of shot-blasting or bush-hammering, or grinding to produce a wonderfully smooth surface, revealing the inner structure of concrete, which has been specially prepared to give a terrazzo- or marble-like appearance.

*Plain concrete flags provide a neat and cost-effective paving for all drives, paths and patios. (Tobermore)*

*Low-cost, budget or economy riven-effect flags tend to have a simple moulding, no colour variation within each flags and come in a limited range of sizes, but can still create good-looking patios. (Brett)*

*Cast Concrete*
Most wet-cast flags are destined for use in homes and gardens. Wet-cast production has one major problem: to faithfully reproduce an intricately detailed mould, a fine-grained concrete is needed, but fine-grains mean limited resistance to abrasion. It is a trade-off between realism and usability. While wet-cast products are fine for low traffic projects (such as patios), the surface detail would quickly be

*Shot-blasting creates a roughened, attractive surface. (Bradstone)*

*The top-of-the-range wet-cast flags often include special features such as circles or octagons. (Brett)*

*Reproduction brick-effect flags bring a traditional look. (Bradstone)*

*Top-quality, wet-cast reproductions are virtually indistinguishable from genuine flagstones. (Marshalls)*

*Setts and cubes are another popular theme for reproducing using a concrete flag. (Stoneflair)*

worn away to base concrete if they were used in a busy civic square or public arena.

Cast flags seek often to imitate some other form of paving. The most common form is the riven effect, which attempts to recreate quarry-fettled stone flags. Given the relatively high cost of natural stone flags (before the arrival of stones from foreign climes), it made economic sense to make concrete copies that could compete on price, if not on authenticity.

The riven-effect flags may be roughly split into three classes. The 'budget' end of the market offers cheap and cheerful copies that really fool no one into thinking that they are anything other than what they really are: concrete copies. Then there is the mid-range, offering a greater degree of realism, better colouring, and a wider range of sizes. These provide reasonable value for money, but there is still little doubt about their real origin. Finally, there are the top-end products, which are uncannily accurate in terms of texture and colour. These can be so realistic that it is not easy for the layman to identify them as concrete.

Other than the riven copies, there are wet-cast flags that seek to imitate old bricks, setts and cobbles, slate, and even timber in the form of decking or railway sleepers. In some cases, these products provide a complex surface in a manageable format and at a reasonable price. In others, they provide a safer alternative to the genuine article, but there are also a significant number of products in this sector that defy explanation and can only be described as decorative and tawdry.

## Stone Flags

For the purposes of description, stone flags may be split into three groups, based on their source, rather than on any idea of rock type or texture. These groups are newly-quarried native stone, imported stone, and reclaimed stone.

**New native stone**  This group comprises flagstones hewn from quarries in the British Isles. Many of these are the famed Yorkstone from the Pennine hills of Yorkshire, Lancashire and Derbyshire. These are primarily gritstones, sandstones and siltstones, deposited as flattish, horizontal beds that can be cleaved or riven into convenient thicknesses and

further split into flagstones. However, flag stone is also quarried elsewhere in Britain: Pennant stone in South Wales, sandstone in the Forest of Dean, Caithness slate in Scotland and Liscannor stone from the west of Ireland. Most modern stone is quarried by machinery and sawn to size. It then undergoes secondary processing to create a texture, and there is a wide range of textures from which to choose. Sawn and honed stone reveals its internal colour and structure, while combing or scratching enhances texture and traction.

**Imported stone**  Britain and Ireland are geologically rich, with a wide range of rock types that are used for a variety of construction purposes, and so stone was traditionally imported only as a last resort, when it was specifically needed for prestige schemes, or when a particular colour, texture or rock type was required. However, during the 1990s the relatively high value of modern paving and hard-landscaping stone, combined with minimal logistical costs and low labour costs in the developing nations of Asia and South America, made it economically viable to bring stone into Britain, Ireland and western Europe. Consequently, we now have a smorgasbord of imported stone. Sandstones from India, similar in colour and texture to the Pennine Yorkstones, are the most popular, but there are also limestones in various shades of grey and blue-grey. Granites from China, travertines from Turkey and porphyry, basalt, quartzite … and so the list goes on.

*Sawn Yorkstone provides a smooth but non-slip surface in colours ranging from bluey-grey to creamy buff yellows.*

**Reclaimed stone** Unlike many other paving materials, stone is almost infinitely reusable. Flagstones that have been pounded by feet and wheels for decades, or even centuries, have an unmistakable character that carries its own premium. The best quality reclaimed flagstones command a price comparable to that of newly quarried stone. Old stone can be sandblasted to bring it up like new, while broken or gnarled corners can be trimmed off to give a whole rectangle once again.

For renovation and heritage projects, and for jobs such as creating cottage-garden-style paving, reclaimed stone offers something not available from newly-quarried materials: a time-worn appearance, so that, even flags laid last week can look as though they have been there since Adam was a lad. Such is the cachet attached to these reclaimed materials that antiquing or distressing techniques have been devised that can be used as secondary processes to convert medium-value, imported sandstone into high-value, worn-weathered stone, adding 60 to 100 per cent to the selling price.

## SETTS, CUBES AND COBBLES

### Definitions

To the average citizen on the Clapham omnibus, they are all cobbled streets, but to a streetmason or paving aficionado, there is a significant difference between setts, cubes and cobbles, and so it makes sense to start by defining which is which.

Setts are blocks of stone, usually 75–200mm wide,

75–200mm deep and random lengths of 75–500mm, although these dimensions are not hard and fast. Actual sizes depend on the type of stone, its workability, and local tradition. They can be almost any hard rock: gritstone, sandstone, limestone, granite, basalt, porphyry or quartzite, depending on what is or was available locally.

Cubes are a more refined version of setts, as suggested by the name. The dimensions are such that the hewn blocks are more or less perfect cubes. 75 × 75 × 75mm or 100 × 100 × 100mm are typical examples, although they may be larger or smaller.

Cobbles are a different animal altogether. They are not hewn from bedrock and they are not cuboid (box-shaped) in form. They are rounded, large pebbles or small boulders, ellipsoid in shape (like a stretched egg); think of the large pebbles found on a beach such as Brighton or Killiney – those rounded lumps of rock that make walking so difficult are cobbles. Their shape is a result of weathering and erosion in a river or the sea, where they are knocked against other rocks, wearing away all the corners and edges, leaving the typical, rounded, egg-shaped cobble. They are readily found in gravel deposits laid down by a river, the sea or as a result of glaciation, and they are found in both upland and lowland areas. That is why they were a popular, cheap surfacing in days gone by. They could be plucked from a river bed, an esker, or a shallow gravel pit and used immediately in an undressed state.

So, that is that cleared up – you can now correct all

*Imported flagstones offer an attractive natural product at an attractive price. (Global Stone)*

*Reclaimed Yorkstone flags have an unmistakable charm that comes only with age.*

*ABOVE: A recent innovation is to mount shallow 'setts' on to a mesh backing and lay them en masse over a prepared bed. (Rock Unique)*

*RIGHT: In this streetscene, the 'path' on the right-hand edge and the channel have been laid using setts, while the main carriageway is paved with cobbles – a true cobbled street.*

the unenlightened when they talk about 'cobbled streets'.

## Construction Methods

There are two methods of construction used with setts, cubes and cobbles: rigid and flexible. For thousands of years, flexible construction was the only option. The units were laid on a bed of loose material, such as sand, gravel, clay or cinders, hammered down to the required level and then the joints would be filled, usually with a granular material, such as more sand, gravel or crushed rock (splitt). This methodology was the basis of many Roman roads, of which several have survived well beyond their design life.

In more recent times, pitch or bitumen was used as a jointing material and is remembered with some fondness as a source of entertainment on the warmer days of summer when it could be extracted with the aid of a lolly stick. As both the jointing and the bedding allowed for minor movement of the pavings, and adjusted themselves to accommodate any such movement while simultaneously maintaining the serviceability of the structure, this method is known as 'flexible' construction.

With the advent of reliable modern cements in the late nineteenth century, concrete bedding and mortar jointing became more popular, and the rigid construction method became the norm as designers and specifiers placed their faith in modern materials, with mixed results, it has to be said. The key difference is that the hard and fast nature of the cementitious bedding and jointing resulted in an inflexible or 'rigid' pavement, which could not accommodate any movement of the pavings and could only respond to external and internal stresses by moving as a whole or, more commonly, cracking along a plane of weakness.

While some rigid pavements have provided years or even decades of service, there have also been a number of dismal failures where a poor understanding of the forces acting within a pavement have

*Sawn setts laid using a 'flexible' construction, the same as is used with concrete block paving. (Marshalls)*

*Pitch-jointing offers a strong contrast with the setts.*

*Riven setts laid using a 'rigid' construction, which features mortar jointing. (Marshalls)*

resulted in loose pavings, collapsed surfaces and an unwarranted assumption that these types of material have had their day. However, intensive research over the last decade, combined with renewed interest in natural stone for paving in the commercial sector, has spurred development of new standards for both rigid and flexible construction methods, and this has resulted in a resurgence in demand for streetmasons with their almost forgotten skills that were used for hundreds of years to lay the types of pavement more familiar to our great-grandparents.

## LOOSE AGGREGATES

These can make a valid claim to be the oldest paving material used by our species. There is nothing particular revolutionary in using a hard, dry, granular material to cover an area of mud or slutch as an aid to traction and a way of keeping the feet drier. Sands or gravels are obvious candidates, but the rich geology of these islands has resulted in a wide variety of materials being used in different parts of the country. Some have found a nationwide appeal, while others remain distinctively local.

### Gravels

Gravels are typically loose, granular materials, ranging in size from 6mm (¼in) to 60mm (2½in). Anything smaller is considered to be 'grit' and anything larger is a cobble. Gravels are divided into two types: the rounded gravels created as part of the natural rock cycle, and angular gravels created as a by-product of quarrying rock.

### Rounded

Rounded gravels are typically formed by moving water: rivers, seas and glaciers. Individual lumps of broken rock are battered by other debris carried by the water, knocking off any rough edges and, in conjunction with other natural forces, breaking apart the rock pieces until they reach a size small enough to be carried by the water, whereupon they roll, bump and bash their way along the river or sea bed in a process referred to as saltation.

At some point, the energy of the water subsides to a point where it is no longer capable of carrying the rock fragments, which by now are smoother and rounder than when they first entered the water, and thus gravel beds are deposited. As the energy of the river dissipates, the current slows and progressively finer particles are deposited. Grits, sands, silts and muds are dropped and left behind as remarkably well-sorted aggregates. Gravels of this type are simply dug direct from the ground, run through a series of sieves to sort them into several size 'envelopes' and packed ready for sale.

Many gravel pits are actually a combination of sands and gravels, as the rivers responsible for deposition varied in energy through the seasons. So, a typical quarrying plant will separate finer sands suitable for use in the preparation of mortars from the coarser sand needed for concretes, and then the gravels of various sizes. The coarse sands are invaluable in many types of segmental pavement construction, as will be seen later.

Rounded gravels from fluvial (river) or marine (sea) sources have the colours and textures of the parent rock, which is usually to be found some way upstream. Only the harder rock types tend to survive the harsh treatment and so chalks and limestones are absent, but granites, quartzites, sandstones and the like are plentiful, and come in an almost unlimited range of colours, textures and lustres.

*Angular*

Angular gravels are most typically formed as a by-product of crushing. Large blocks of rock are fed into immensely powerful crushing machines that break up the rock to more manageable sizes that are subsequently screened and blended to form construction aggregates. The gravels are usually described as 'clean', which means that they are solid lumps of rock with few or no fines (dust or sandlike particles). Angular gravels give a greater degree of interlock between particles, compared with the rounded gravels, and so tend to be more stable.

*Self-binding Gravels*

Gravels make a perfectly acceptable surface dressing, but, when laid to a depth of more than around twice that of the grade size, they become difficult for traffic since feet and wheels tend to sink into them, rather than flit across. This phenomenon is used to good effect with the gravel traps seen alongside

steeply descending roads or at racetracks. A vehicle entering the gravel trap finds that the material moves when loaded and is incapable of supporting the vehicle's weight which then gradually sinks into the gravel and comes to a safe and, one hopes, dignified halt.

This may be a problem when gravel is required to be used for a pathway or driveway. The tendency of vehicles and foot traffic to sink into it is a definite disadvantage in such situations, and so carefully considered construction methods have to be used, or some way of 'binding' the gravels to enable them to support heavier loads will be required. The simplest way of achieving this binding is to use a gravel that contains a quantity of fines which fill the voids between individual stones and so reduce the opportunity for movement. These gravels are referred to as self-binding and there are numerous types found throughout the land. Possibly the most famous is the Breedon gravel from the borders of Nottinghamshire and Derbyshire. It is essentially a crushed limestone, but its particular characteristics and its distinctive, warm, buff colouring have made it a favourite in parks and gardens looking for a low-cost yet effective and natural looking surfacing. It is a big favourite of Queen Elizabeth II, who has used it on a number of Crown properties.

ABOVE: *Contrast between rounded and angular gravels. (Border Stone)*

RIGHT: *Gravel paths can be mixed and matched with many other natural materials, such as these stepping stones. (Stoneflair)*

31

Other self-binding gravels include Cedec, a proprietary brand manufactured from crushed granite, and Coxwell gravel from Oxfordshire, which is a brown-buff blend from the Cherwell Valley. Local, limestone-based self-binding gravels are common throughout those parts of the country with a limestone geology, and the same product may be known by different names in adjacent towns. There are also sandstone-based and granite-based self-binding gravels.

*Hoggin*

This is a classic example of a local or regional aggregate. Ask anyone north of the Watford Gap and there is a good chance that they will scratch their head and wonder what you are talking about. Hoggin is a product of south-east England, an area that is, relative to the rest of these islands, geologically poor. The predominance of chalks, greensands and clays means there are not many rock types suitable for use as loose aggregates, so when a self-binding gravel is required, the locals have taken to using a naturally existing mixture of clays, sand and gravels, which they call 'hoggin'. It is similar in many respects to the self-binding gravels of further north, but the clay content often renders it sticky and claggy in wet weather. However, when you have no adequate alternative it is better than nothing. (Probably.)

*Other Loose Materials*

Recent legislation has imposed a system of progressive taxation on newly quarried aggregates (referred to as 'primary aggregates'), and a market for recycled or reclaimed aggregates is beginning to develop. These range from the exotic, such as crushed and tumbled coloured glass, to the surprising, such as crushed brick, to the downright ordinary, such as crushed concrete.

There are also woodchips, formed from trashed timber pallets, available in a range of colours from natural to browns to the quite outlandish – how much demand can there be for vivid purple woodchips?

And there is crag, a local name used in East Anglia for crushed seashells. This product has found favour among horse lovers as the shells are said to be soft and kinder to the hooves of their steeds. Also popular for equestrian purposes is recycled crumb rubber and a specific 'arena sand'.

## Cellular Systems

A modern approach to binding or containing gravels is to use cell matrix systems. These can be thought of as interlocking, bottomless, plastic trays that consist of a number of cells roughly 50 × 50mm (2 × 2in) that are filled with the chosen gravel. The cell matrices come in all sorts of format: squares, hexagons, octagons and unnameable intricate shapes, but the end result is a stabilized gravel. The underlying cell structure is just about visible, but the overall effect is quite naturalistic. The cells are also used with grass; laid on a free-draining substrate and filled with

*A self-binding gravel is less prone to scattering and is a good choice for both garden paths and driveways.*

*Crushed glass aggregates bring vivid colours and textures to a garden, especially when combined with strong contrast features, such as these rainbow cobbles. (Brett)*

*Cellular matrix products can stabilize gravels, but may also be filled with grass for occasional use. (Aco)*

*Plain concrete is easy to install and cheap, but it is not the most interesting surface to look at.*

a specially selected sandy loam they provide sufficient strength to carry quite substantial loads while retaining the look of a lawned or grassed area. However, although they cope well with occasional traffic (hence their use for fire lanes and the like), the grass cannot cope with regular, day-to-day traffic and so grass-filled cells are not really suitable for driveway use.

## PLAIN AND PATTERNED CONCRETE

### Plain Concrete

Concrete is usually regarded as a cheap, plain and simple surfacing that is not particularly attractive. It is best described as 'utilitarian'. It does what it is supposed to do, provide a firm and safe footing, and nothing more. It was popular during the 1970s as car ownership grew and homes old and new needed a cheap and effective surface that could take the weight of a car without bankrupting the owners.

Things have moved on since then and there are more decorative options available to us now. However, it would be a mistake to underestimate the usefulness of plain, straightforward, no-nonsense concrete as a hardstanding for sheds and garages, or combined with a more decorative edging to form low-cost paths in less aesthetically demanding areas of the home and garden.

The basic plain concrete hardstanding has a number of finishes. Driveways and paths with a gradient would typically be given a 'tamp' finish,

created by levelling the wet concrete by using a plank or something similar, parallel to one end, and leaving a series of shallow ridges running across the surface (transverse) and intended to improve traction. Decorative areas would be trowelled, but this left a smooth surface that was less than ideal for outdoors in winter, and so these areas would normally be 'brush-and-trowel finished'. The smooth-trowelled surface would be roughened by drawing a brush across the surface in one direction, to create small mini-ridges in the surface, and then float-finish the edges into a smooth boundary strip. The more artistically inclined would be tempted to impart a crazy-paved effect finish by marking out irregular, interlocked polygons into the freshly trowelled surface.

Colour may be added to a concrete – it doesn't have to be that uninspiring shade of flat grey. It can be red or yellow, brown or green. Many different colours are available by incorporating dyes into the mix, but it should be borne in mind that even the best quality dyes fade in UV light over time, and what might be a deep red this year will probably be a shade of pink in five or ten years' time.

### PIC and Decorative Concrete

The decorative effects of imprinting a pattern and adding colour have been combined in the technique known as pattern imprinted concrete (PIC), also known as stamped concrete. This technique has gained in popularity over the last few years, but it is not to everyone's taste and it can be relatively

*Although popular for use on driveways, there is no reason why pattern imprinted concrete cannot be used for patios and garden features. (Creative Impressions)*

*Acid-stained concrete is more popular for internal flooring, particularly for conservatories where it forms a 'halfway house' between home and garden. (Creative Impressions)*

expensive for smaller areas. The basic premise is that the surface of the concrete is treated with "colour hardener', which is a high-strength, cement-based, coloured powder, scattered on to the fresh concrete, trowelled in to bring out the chosen colour, and then 'stamped' with textured mats to imprint a pattern. The patterns available are legion, but the most popular tend to be those that emulate old setts, riven stone, cobbles or some other surface that would be much more difficult (and usually even more expensive) to install.

The completed surface is treated with a sealant to guard it against accidental staining and also to enhance the colouring while simultaneously protecting it from the bleaching effect of UV. The installers often claim that this type of surfacing is 'maintenance-free', but it will still need an occasional brushing to get rid of the detritus that builds up on

any surface, and it is quite likely that the sealant will need to be reapplied every couple of years or so, if the pavement is to maintain its looks.

### Acid-stain

Other techniques for introducing colour to concrete include acid-staining. This is not widely used in Britain and Ireland, but is more common in the USA, where it is something of an art form. By applying certain minerals and acids to the surface of a concrete, the chemistry can be altered and beguiling, natural-looking colours with an organic variation in shade can be created. The technique is rather hit-and-miss, and needs to be done by a competent contractor familiar with the process and capable of bringing out the best. Using too much or too little of the stains can have serious consequences for the finished appearance.

### Exposed Aggregate

A more established method of introducing colour and texture is to employ an exposed aggregate technique. This simple idea uses an attractively coloured aggregate on the surface of the concrete to give it a more appealing look. There are two principle methods that can be used: the decorative aggregate may be sprinkled over the surface of the fresh concrete and tapped down into the matrix, or it can be incorporated within the concrete and then treated with a special sugar-based liquid that prevents setting in the uppermost few millimetres of cement. A few

*An exposed aggregate finish gives concrete a new look at little extra cost.*

hours later the surface can be hosed down with a power washer, which washes away that uppermost, unset matrix, revealing the decorative aggregate held firmly in place by the unaffected concrete matrix beneath. So, when someone mentions concrete, do not assume it has to be the plain and boring surface of forty years ago. There is a whole new world of concrete, full of colour and texture limited only by your imagination.

## TARMAC

Tarmac, or bituminous macadam (bitmac) to give its full name, has been with us for over a century but has fallen out of favour for residential paths and driveways in recent years, as property owners have looked to surfaces offering more colour, texture and style. It is still used for driveways to new-build houses since it is cheap and quick when laid in quantity, but for a one-off job, such as a typical driveway reconstruction, it may be quite expensive mainly because of the amount of specialist equipment and the number of operatives needed to lay it.

Most people assume that tarmac comes in the one colour – black, but a red tarmac has been used for a more upmarket look for many years and it is now possible to have almost any colour you want, thanks to recent advances in binder technology. Even among the more common colours there is a degree of variation in quality and some care is needed when specifying a bitmac pavement to ensure consistency.

Bitmac consist of two main ingredients: the binder, which once, long ago, was tar, but is now most commonly a modified bitumen, and the aggregate. A basic black bitmac will use the standard black binder with a cheap local aggregate, often limestone. When first laid, the binder covers all of the aggregate and you have a wonderful, jet-black surface. However, the binder soon wears off the surface to reveal the aggregate beneath. Where a limestone has been used, this becomes the predominant colour, which explains why some bitmac pavements look grey, or almost white, after just a few years. A better quality bitmac will use a dark or black hardstone as the aggregate, so that even when the thin smear of binder covering the surface has been worn away, the pavement still looks black.

*Red macadam is regarded as more upmarket than the more traditional black variety, even though the price difference is only a few pounds per tonne.*

With the red bitmacs this principle is even more important, because of the premium charged for a red macadam surface. With the 'budget' red bitmac, the binder will be coloured red, but the aggregate may be a limestone, and, as soon as that smear of red binder is abraded by traffic, the surface takes on the colour of the aggregate. The better quality red bitmac materials use red quartzite as the aggregate so that, again, even after years of wear, the surface remains the intended colour.

The other variable factor with bitmac is texture and this, too, is determined by the aggregate content. A bitmac driveway or path should consist of at least two separate layers of macadam. A coarser, chunkier base course, now termed the 'binder course', that gives strength to the pavement, and a finer, more accurately laid 'surface course' that is intended to give a smooth, neat and attractive-looking finish to the pavement.

*Most tarmac is black and a small, 6mm surface course gives a better-looking finish for residential projects.*

The surface course will typically have an 'open' or a 'dense' texture. Open texture is similar to that of the Rice Krispie cakes made by schoolchildren: regular sized aggregates stuck together with a binder, while a dense texture uses a range of particle sizes, to give a closer, fuller, less holey texture. For most residential surfacing, a dense textured finish is more appropriate. The size of the aggregate will also affect the appearance. A 10mm aggregate will look quite 'bitty' and is better suited to public footpaths and carriageways. A 6mm aggregate gives the tighter, smoother finish that is popular with homeowners.

For a really smooth finish, an asphalt or stone mastic asphalt (SMA) is preferred. These use very fine gravels and sands in the macadam and so a very 'tight' finish can be achieved. Chippings of one form or another are often rolled into the surface to provide a harder, tougher running surface, and to improve traction. This type of surfacing is popular for many of our major roads and motorways, but may be used to good effect for residential driveways, if a competent contractor can be found.

It should be pointed out that bitmac surfacing is one of those tasks that really are best left to the experts. It uses a number of specialist tools that are not likely to be found in the average DIY-er's shed, and it has to be worked fast and accurately as the macadam is cooling all the time, and once it becomes too cool, it is effectively unworkable.

## RESIN-BASED

Resins comprise another technology that has mushroomed over the last twenty years or so. Originally developed as glues, they went on to find a great many applications in the wider construction industry, and farther afield. As far as paving is concerned, their principal application is as a binder to affix decorative aggregates to a new or an existing substrate, which should be a monolithic or composite surface (that is, a solid mass of paving) rather than a modular or segmental surface (one composed of a number of individual elements, such as flags, setts or block pavers). Concrete or tarmac surfaces are usually fine, as long as they are in reasonably good condition.

The finished effect is similar to that created with the exposed aggregate concretes described previously, although the more gravelly appearance is, perhaps, more akin to that of a self-binding gravel. Obviously, the type and the size of the aggregate chosen determines just how the surfacing will appear, but the exact type of system chosen also has a significant effect on the precise look.

There are two principal methods: in the first,

*Resin-fixed aggregates give a traditional look with none of the problems normally associated with loose gravels. (DecorDrive)*

*Resin-bound, natural gravels can even be used with quite considerable gradients. (SureSet)*

*By using a resin-bound system, materials such as crushed and tumbled glass can be used to create highly original features. (SureSet)*

known as a bonded surface, the resin is applied to the substrate and then the selected gravel is scattered over the sticky resin to create an even coating. The second method relies on mixing the selected gravel with the resin and then trowelling it over the substrate. This is known as a resin-bound surface. The similarity in names does not greatly help the general public to understand the differences between them, and so the more descriptive names of 'scattercoat' and 'trowelled' will be used henceforth.

Scattercoat systems are generally cheaper than trowelled systems and are more suitable for DIY. The finished texture is quite granular, since it takes on the actual surface profile of the selected aggregate. There is a definite crunchy quality to the completed surface, emphasizing the gravel heritage of the surface, but it has to be said that some of the products are not as good as they could be and some of the aggregates supplied are far too big. The better manufacturers and installers guarantee their products and tend to supply bauxite or similar aggregates of 6mm size or smaller.

By contrast, trowelled systems give a noticeably smoother finish, with no 'crunch' when subjected to traffic. To achieve this smooth, even finish, an experienced installer is highly recommended since these systems are not really suitable for DIY applications. However, because the resin coats all of the aggregate, and not just the substrate, a much wider range of aggregates may be used, including larger sizes, up to 20mm or so, along with intriguing materials such as, glass beads or crushed CDs.

Whichever technique is chosen, the result is a natural looking, low-maintenance and hard-wearing finish, with superb design and detailing opportunities. In the residential market these products are mainly targeted at driveways, but they are eminently suitable for paths and patios, and resin-bound trowelled surfaces can even be extended into the home, providing a themed link between the internal and the external environment.

## DECKING

Decking has been a popular garden surface for the last few years, thanks mainly to the efforts of television make-over shows. How long its popularity can

*Outside a conservatory, a deck can make a pleasant transition between house and garden.*

be sustained is not certain since it does seem to be the whims of fashion that sustain its sales rather than any particular structural advantages.

Timber is never suitable for use as a driveway, but some people like to use it for paths and particularly for terraces and patios. It has been used to great effect in warmer and drier climes, but in the cool and damp atmosphere of the British Isles, even treated timber struggles to cope with mosses and algae, and many decks need intensive, regular maintenance to keep them not just clean and serviceable, but safe and sound.

A basic patio deck is relatively simple to construct and does not require much of the digging and preparatory work needed for other surfaces. It is a task that is perfectly suited to the home DIY-er, and many suppliers provide the materials in kit form that need little more than nailing and screwing together to create a deck designed to complement most modern homes. The cost also helps to make decking an attractive option since it compares well with even the cheapest hard-paving alternative and can be only a fraction of the cost of a high-specification surface, such as tumbled block paving or natural stone flags.

However, the inevitable slipperiness of damp timber and the critical need for annual maintenance should not be overlooked nor underestimated. Many decking owners have reluctantly resorted to covering their timbers with chicken wire or a high-grip sealant in an attempt to render the surface safer for children, the disabled and those of us less agile on their feet.

# CHAPTER 3

# The Design Process

## INTRODUCTION

How is a patio or driveway designed? Is it simply a matter of choosing a surface that you like and fitting it into the area available, or is there more to it than that? What needs to be considered before implementing a plan?

Obviously, choosing a type of paving that suits one's taste is the first criterion to be considered. On

*OPPOSITE: Circle features or circular layouts change the dynamics of a pavement by drawing the eye to specific points. (Marshalls)*

any project there should be one main form of paving, although this may be accented by the use of a contrast. The main paving has to be something that suits the taste and sensitivities of the homeowner, something that complements the property and the local area, and, of course, something that is not going to break the budget.

There are no hard and fast rules that must be followed, no flow chart that can lead to the right paving for any given project. If it were that simple, every housing estate would feature homes with identical driveways – the fact that some housing estates *do* feature the same driveway over and over again serves only to show that there is a distinct lack of design consideration used in some cases.

---

### Units of Measurement

Although yards, feet, inches, gallons and pounds are still used in everyday discussions, the paving industry has been more or less metric since the mid 1970s, and so metric units have been used throughout this book. However, as an aid to understanding, a basic table of comparisons might be useful for those unfamiliar with millimetres, kilograms and litres.

**Length**
25mm ≈ 1in
50mm ≈ 2in
75mm ≈ 3in
100mm ≈ 4in
150mm ≈ 6in
200mm ≈ 8in
300mm ≈ 9in
450mm ≈ 18in or 1ft 6in
600mm ≈ 24in or 2ft
900mm ≈ 36in or 3ft or 1yd
1000mm = 1m ≈ 39½in

**Mass**
100g ≈ 4oz
1,000g = 1kg ≈ 40oz ≈ 2.2lb
25kg ≈ 56lb = ½cwt
50kg ≈ 112lb = 1cwt
1000kg = 1 tonne ≈ 2,204lb

**Area**
$1m^2$ ≈ 1.2sq.yd
**Volume**
$1m^3$ ≈ 1.3cu.yd

---

*Highmoor banded Yorkstone is one of the most beautiful and distinctive flagstones in the world.*

*The use of a red block edging gives a 'designed' look to even a simple tarmac driveway.*

## A MATERIALS SHORTLIST

Possibly the simplest design methodology is to eliminate those types of paving that are definitely off the menu. If the house is a cutting-edge modern design, will reclaimed Yorkstone really be suitable? Similarly, is pattern imprinted concrete really the best idea for an Edwardian family home? Or red 200 × 100mm block pavers for a Cotswold cottage?

Try to narrow down the overwhelming choice to two, but no more than three, types of paving. At this early stage, there is no need to focus on specific items, general types will suffice, so decide whether the main paving is to be block paving, tarmac, setts or whatever, and then the rest can be disregarded.

On larger projects it might not be practicable to use the same main paving throughout. If bitmac has been selected as the surfacing for the driveway, it is unlikely to be appropriate for the patio; however it is a simple matter to use one consistent theme to link patio to path and path to driveway. Maybe the same edging is used throughout or a motif, such as a diamond-shaped panel of pavers, is regularly repeated through the project.

Concentrate on the main paving for now. When others see the completed work will they recall it as a flagged area, a concreted area, setts? Identifying the main paving type, or at least reducing the field to just a couple of options, will help to clear the mind and the drawing board can be prepared for the next step

of the design process, which is often the choice of a complementary or contrasting paving to use in conjunction with the main surface.

### Using Contrast

Consider a paving as beautiful as a Highmoor Yorkstone. Each and every flagstone or sett is unique and a natural work of art, yet if it were laid over a large area, say 10m × 10m, those individually attractive stones become a meaningless, almost featureless mass. There is too much, and it is repeated. What is needed is some contrast, something similar yet distinct, something that will emphasize the natural beauty of the stone. It may be a contrasting colour or texture in the same material, a plainer flagstone, perhaps a Crosland Hill with its creamy-buff colouring, or it might be a completely different type of stone, perhaps a dark Welsh slate or a sparkly Cornish granite, and the difference might be accentuated by using a different format – using setts as contrast to the flagstones. We could take it further and use a different material, perhaps a clay paver or even a chromed edging detail.

The same principles apply with manufactured pavers: consider a tumbled paver. Again, it looks attractive on a 3m wide single driveway, but if we upgrade to a 5.5m double-width driveway, those individually attractive blocks become lost in the ocean. So we introduce a contrasting edge course – maybe the same blocks turned through 90 degrees and laid as parallel courses, or a block of a contrasting colour, maybe even a smaller or a larger block to emphasize the contrast, or maybe we shall go all the way and use a contrasting size, orientation *and* colour.

With a plainer surface, such as bitmac or concrete, having an expanse of the same, monochromatic finish can and does look boring. That is why edgings are such a popular choice with these 'simpler' surfacings. Adding a 200mm-wide edge course of concrete block pavers in, say, a red colour, can magically transform a featureless spread into a driveway with individual character, and make it look as though someone had put some thought into it and not gone for the simplest, easiest solution.

## Mix and Match

So, contrast can be used to emphasize the detail and aesthetics of the main surface, and materials can be mixed and matched to suit the needs of the project; but are there any limitations? Not all materials work well together. As a generalization, natural materials tend to sit uncomfortably with man-made. It is not a hard-and-fast rule, it is a matter of aesthetics and one person's Canaletto is another's Jackson Pollock. If the materials suit the eye of the beholder, then that is all that matters.

It is something to do with colours and textures – natural colours have organic, muted tones, and authentic, subtle textures. These are often mimicked by manufactured products, to varying degrees of success, but even the very best man-made products never quite manage that organic touch of originality. On some projects, the haphazard styling and variable colouring of natural products may not be suitable, and the uniform colours and regular texturing of man-made materials may blend better with the buildings and environment. However, it is when natural and man-made materials of a similar format are combined that the juxtaposition may be awkward.

Natural stone flags edged with concrete block pavers look uncomfortable, but if a battered, tumbled concrete paver is used, the mismatch is eased. If a clay or a tumbled clay paver is used, then the materials cooperate rather than clash. Similarly, clay and concrete pavers do not mix – the strong natural colours of the clays totally overwhelm and actually 'bleach' the manufactured colours of the concrete pavers to the eye.

When choosing a contrast paving, for the edging or for detailing, choosing a material from the same group often (but not always) produces a more sympathetic combination. Manufactured with manufactured; natural with natural, but do not be afraid to experiment. A further consideration when choosing a contrast paving is the construction method. If the design requires concrete pavers to be mixed with concrete flags, it is possible that both blocks and flags can go on the same bed and so simplify the construction. This happy coincidence also has the benefit of simplifying the installation of features such as circles or diamonds within the main paving. With other

*The clay paver, on the left, has a stronger and more subtle colouring than the concrete equivalent on the right.*

materials, two stages of construction may be necessary, and this can complicate the construction process, particularly when a contrasting material is to be used within the main paving.

## DESIGN CONSIDERATIONS

### Defining the Function

Once the choice of materials has been made, or the field narrowed down considerably, the next step is to determine just how the pavement(s) will be used. This may seem obvious, but by thinking it through, many subsidiary questions come to mind and the finished design is much more likely to be in keeping with the needs of the project, and the limitations of the wallet.

What is meant by 'defining the function'? It refers to the process of thinking about how the pavement will be used and ensuring that the form will suit the function. If this is to be a patio for a young family, granite setts will not be the best choice, even if the house is a period property and granite setts are the traditional local material. Young families have children with bicycles, prams and ball games: granite setts are uneven and have relatively wide joints that trap small wheels and cause erratic bouncing of footballs and basketballs. Do you want to spend valuable leisure time giving the children a push to get them moving, or replacing window glass lost to a ball that seemed to have a will of its own?

Is the paving a decorative area, or an area for socializing, eating and drinking? Is it just a parking area, or is it a combined driveway, parking area and main access to the home? Will it be used every day, every week, or is it seasonal? Will it be trafficked by feet and wheels on a regular basis, or will there be a caravan or a boat parked on it for long periods? How will it be cleaned and maintained? If it is a yard where dogs are kept, how will the surfacing and any jointing hold up to the particular demands made by pets?

Choosing the paving that appeals most and throwing it down in any given area is not always a great success. Those lovely, riven-effect flags seen in the garden centre are not actually suitable for use on a driveway, and do you really want to pay £70 or more per square metre for paving that will be permanently hidden underneath a shed, summerhouse or caravan? By thinking about how the pavement will be used, the choice of materials, the construction method, the detailing, the drainage, and a host of other factors are brought into focus.

### Kerbs, Edgings and Mowing Strips

Will these be required? With some forms of paving, a fixed edge of some sort is absolutely essential; with others, it is purely decorative, but can be the finishing touch that completes the image. What is the difference between a kerb and an edging? Maybe there is no official definition, but a useful way to think about them is that a kerb provides check – upstand that prevents users from straying beyond the edge; an edging, on the other hand, is purely a restraining

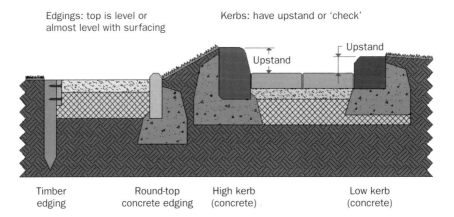

*Contrast between kerbs and edgings: kerbs have upstand (check), whereas edgings are flush with the surface.*

A selection of decorative edging kerbs. (Bradstone)

ABOVE: *Mowing strips can help to simplify basic lawn maintenance as well as adding a decorative feature to a garden. (Bradstone)*

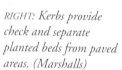

RIGHT: *Kerbs provide check and separate planted beds from paved areas. (Marshalls)*

construction that lies flush with, or slightly proud of, the pavement. Just to confuse matters, the features referred to as kerbs can be used in ways that result in no upstand, so they are, technically, edgings, and some items sold as 'edgings' can be laid with upstand, so they could be kerbs.

The primary function of any kerb or edging should be to restrain the pavement. As indicated earlier, the pavement might not actually need restraining, but the edging should seek to give the impression that, without it, the pavement would disappear into the garden. As a restraining structure,

an edging must be firmly and soundly constructed, which means using concrete or mortar to hold everything in place. The use of concrete brings problems of its own. There needs to be space to accommodate the concrete haunching and then there needs to be some way of making the haunching hidden or discreet. Where a planted or lawned area lies behind the kerb or edging, the haunching needs to be kept lower than the top of the edging so that it can be buried with soil or turf. However, sufficient haunching needs to be used to ensure that the edging is capable of performing its primary function.

*Patios need plenty of room. (Tobermore)*

Some edging products are manufactured for discretion, some for show and some for solidity. Part of the design process involves deciding, first of all, whether edgings or kerbs are required, and, if they are, what type is most suitable? Kerbs are often a better choice for those areas where vehicles are likely, since they provide that essential check that is designed to keep cars, cycles and other wheeled vehicles on the paved surfaces and off the gardens or footpaths. They are also useful with planter beds, as they enable the soil level to be elevated, which increases the soil depth and aids drainage, all of which benefit the plants.

For lawned areas, decorative areas, or those set aside for play or socializing, the presence of a kerb may be more of a nuisance than a benefit. With lawns, it is impossible to get the mower tight up to the kerb face when cutting the grass, so the gardener is forced to return later with the strimmer to tidy up these edges. Where an edging has been used, the mower should be able to pass over both lawn and edging in one pass and eliminate the need for additional strimming.

One option to consider if there is no alternative other than to use a kerb, or for those areas where a lawn lies directly against some other vertical structure, such as a wall, is to use a 'mowing strip'. This is simply a narrow strip of paving laid flush with the lawn and tight against the kerb or wall; it acts to transfer the edge of the lawn away from the vertical face, thereby allowing a mower to pass straight over, as described above. This option combines all the advantages of both kerbs and flush edgings: the provision of check to keep vehicles and the like in their place, while simplifying lawn maintenance.

## Size and Shape

This part of the design process is easily overlooked. There is a great temptation to fill an area with paving without giving any thought as to whether this is actually the best option. One common request encountered by paving contractors when discussing options with homeowners is to pave the entire frontage of the house – do away with all the lawn, all the planting and pave the lot. This might seem like a simple option but the finished effect is usually disappointing. All that is needed is a couple of fuel pumps and you have the perfect petrol station look. Similarly with rear gardens, especially the smaller ones – flooding the whole area with hard paving may actually make the space seem smaller by linking the hard surface to the vertical boundaries and thereby tricking the eye into drawing them inwards.

Conversely, with some layouts, the space allocated

to hard paving is parsimonious, to say the least, and the result is the 'new house' effect, where the driveway is just wide enough to take a car, but the driver and passengers are obliged to step out on to grass, or the rear patio is just large enough to accommodate the patio table but it is impossible to get around the table to other parts of the garden, or to serve food without stepping on to the flower beds or the lawn.

This consideration of size links back to the earlier mention of thinking about the function of the paving. A driveway needs to be large enough for both cars *and* their users; a patio has to be large enough for both furniture *and* people, and a path has to be wide enough to prevent the illusion of falling off its edge.

So what role is played by shape? This relates to the consideration of function and size but also of the type of materials being used. With a modern design, a strong, regular, geometric shape is a common feature, and with manufactured or dimensional materials it may be possible to use shapes and sizes that minimize or eliminate the need for cutting. More traditional designs, particularly those using reclaimed or reproduction materials, often make the 'best use' of materials and so follow less strict, more 'organic' shapes.

Obviously, the shape, size and layout of the property in relation to the site will determine what is possible, but each type of project involves other considerations too. For driveways, the prime consideration has to be size, ensuring that there is sufficient space for the family vehicles, along with room to get in and out of them, and permit easy access to any garage, to a turning head (if one is present) and, of course, egress on to the public highway with ease and good visibility. There may also need to be linked pathways leading to the rear garden or other parts of the property. How these are laid out is the key to successful design. Will it be sweeps and arcs, squares and rectangles, or a looser, unstructured approach?

Patios present different problems. Size is not always the key consideration since a patio does not usually cover half or more of the space available, so the issue of 'balance' often matters more – the way that the needs of the family are balanced with the limitations of space and budget, the need to balance the scale of the hard-landscaping with the rest of the garden, the so-called softscape, and the need to

balance what is, after all, a hard and definitely man-made surface with the natural, forms and colours and textures of the garden.

There needs to be space for any furniture: a table, along with its chairs, and then there may be additional items, such as a patio heater, loungers and awnings. There needs to be circulation space – room enough for people and pets to get from the house to the garden, with simple, clear access from doorways to pathways. But most patios are not open areas of paving, they are often adorned with pots of plants, perhaps with statuary or some other ornamentation and so some thought has to given while planning to how all these different demands will be balanced.

And, of course, the type of material chosen for the paving has an impact on shape and size. Monolithic surfaces, such as concrete, bitmac or resin, can be laid to any and every shape, and with the smaller paving units, such as blocks, cutting to shape is not a big issue, but with larger units, especially the flagstones, cutting curves and arcs becomes problematic.

With smaller elements, such as block and setts, large expanses can be broken up by incorporating 'design features', such as diamonds, circles or other motifs. These help to prevent the eye from being lost within the larger area by creating a focal point that attracts the eye and so draws in the edges, linking them into the whole and giving the impression of a much more manageable space.

The ideal size and shape of the paved area emerges by considering all these factors, as well as the aesthetic factors, but there is no right or wrong, just good and better. The aim of design is to ensure that the finished project is not just good, but is definitely better.

## Driveway Sizes

### Parking Widths

It was mentioned earlier that many driveways on new homes are somewhat narrow. This is not simply penny-pinching on the part of the builders and developers, but stems from a sense of scale and balance, and the need to provide that oft-neglected expanse of shabby turf referred to as the 'front garden'.

A single garage has an entrance width of around

*Driveways to new properties are often excessively narrow.*

*This driveway has been widened into the lawn area to make it easier to get out of the car. (Formpave)*

1.9m. This is generally just wide enough to allow the typical family car to get in and out without damaging the paintwork. The average driveway outside the average garage is set out at a width of 2.4m, usually centred, more or less, on the garage. This is just wide enough to allow a car to be parked and the driver to get out without stepping on to the garden. Any passengers on the nearside are not so lucky.

As many contractors will attest, one of the most popular requests from new estates is to widen the driveway – is there any way to add half a metre or so to one side, so that both the driver and the passengers will have a firm footing when getting out of the car? Have a look around any new estate with 'single' driveways; how many of them have been DIY-widened by adding a row of flags to one or both sides?

Surely it is better with the type of home that has a double garage? If only it were so. The typical double garage has a threshold width of 4.3m. Assuming a

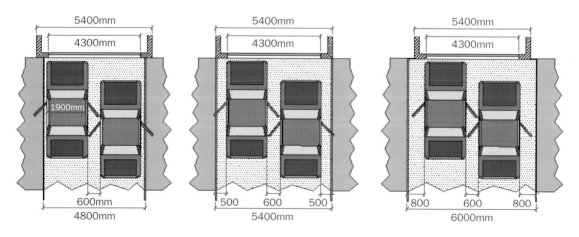

*A typical double drive layout. The standard width is too narrow, while making it the same width as a standard double garage does not improve the situation much. Allowing 3m for each car gives ample room for entering/leaving the vehicles.*

typical width of 1.8m for a European car (excluding wing mirrors), that is just enough room for two cars (3.6m) with a little breathing space. Outside, the driveway is likely to be 4.8m wide (about 16ft). This gives space for the two cars, plus two access channels at 600mm wide. But we need at least three access channels: one to the nearside of the car on the left, one to the off-side of the car on the right and one between the two cars. So, once again, we have the problem of a drive that is too narrow for comfort.

When planning a new driveway for the typical family car, it is always safer to allow 1.8m for the car, then a minimum of 0.6m to each side. That is 3m per car – 6m for a double driveway. It is possible to get away with less, but these figures are a comfortable guide. For the larger car, what is now called an 'executive' or 'luxury' model, the car width may be 2.4m, and some imported models are even wider, so, when planning a driveway, to know the types of vehicle that will be using it is critically important.

*Turning Circles*

Width is not everything: on larger driveways and for trackways, turning circles often need to be considered. The turning circle is a measure of how sharply a vehicle can manoeuvre. Some vehicles, notably black cab taxis, have superlative turning circles – they are often said to be able to 'turn on a sixpence'. Even the

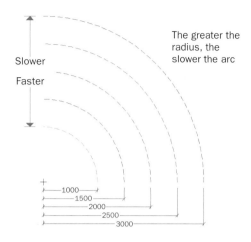

*Faster and slower arcs.*

typical family car or a small van has a reasonable turning circle that allows the vehicle to be manoeuvred with a degree of ease. The turning circle varies from car to car, but usually has a diameter of about 10m; obviously, smaller cars have tighter turning circles and bigger cars have larger circles.

Such factors dictate just how sharply curved a turn or an arc can be. In the construction industry, arcs are described as 'fast' for those with a short radius and 'slow' for those with a longer radius. An arc with a

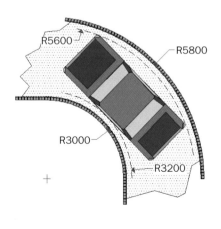

*Turning arcs for a typical family car.*

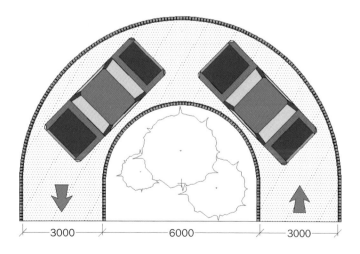

*Minimum recommended layout for a horseshoe drive.*

radius of 4.5m may be described as faster than a 6m arc, but slower than an arc of 3m radius.

Taking the average turning circle diameter of 10m, this gives a radius of 5m. Allowing 1.8m for the average width of the car, this results in the radius of the turning circle on the inside of the vehicle being: 5m – 1.8m = 3.2m. So, in theory, a typical car should be able to navigate an arc of 3.2m. Allowing for driver error and a little room for comfort, it can be seen that an external arc of 3m radius is about as 'fast' as can be negotiated in a single pass by the average family car. The minimum negotiable radius for an internal arc is equal to that of the turning circle data, that is, 5m, although some allowance needs to be made so that full lock is not required and tyres are not rubbing against kerbs; so a minimum internal arc is usually around 5.8m. Thus, when designing a driveway, try to ensure that all arcs have internal radii of not less than 3m and, for external arcs, a minimum radius of 5.8m.

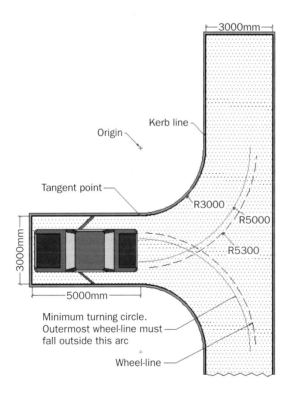

*Minimum recommended layout for a turning head.*

*Horseshoe Layout*

Many residential driveways have one major disadvantage: they are not very big and so vehicles need to be reversed in or reversed out because there simply is not sufficient room to allow for a vehicle to be turned through 180 degrees. This is sometimes overcome by constructing a 'horseshoe' driveway, with a separate entrance and exit to enable vehicles to be driven in and driven out. Larger driveways may have sufficient room to enable a vehicle to complete a 180 degrees turn, and may even incorporate a circle or circular feature to facilitate the manoeuvre. The guidance given above with regard to turning circles indicates that a horseshoe driveway would need to have an inside arc of at least 3m, and, if there is to be a drive width of 3m, as recommended in the section on size and shape, it can be seen that a typical horseshoe or circle layout would be at least 12m in width.

*Turning Heads*

An alternative layout that may also eliminate the need to either reverse in or reverse out of the driveway from the public highway uses what is known as a 'turning head'. This is an area within the driveway laid out in such a way as to allow a vehicle to be driven into one 'arm' of the turning head, reversed into the other 'arm', and then driven out again, having turned through 180 degrees. The great advantage of such a layout is that it transfers the reversing manoeuvre from the risky public highway to the relative safety of the private driveway.

Again, the turning circle data may be used to determine the space required to include a turning head in a driveway. However, it is not just the turning circle of the vehicles that needs to be considered, the length of the vehicle also has to be borne in mind to ensure that there is adequate length in the 'arms' to allow the vehicle to be manoeuvred successfully. A typical layout is shown in the illustration.

## SLOPES, FALLS AND GRADIENTS

Another essential design consideration is drainage. Although there are now permeable and porous paving products that allow water to pass straight through and so do away with the need for falls, gullies and all the other fittings normally used to keep the paved surface above water, the most popular paving materials for residential driveways, paths and patios remain impermeable (more or less) and so some provision for drainage needs to be included in any design.

## Why Drainage Matters

Why do we need to drain paving? There are a number of reasons, but two of the most important are safety and structural integrity. A wet or submerged pavement is dangerous. It is slippery and in winter it will be icy. Given that one of the primary functions of any paving is to provide a safe and firm footing, being underwater, even if it is just a temporary phenomenon, makes no sense. From a structural viewpoint, builders and streetmasons have known since Roman times and earlier that, to maintain the structural integrity of a pavement, the surface water must be removed as quickly and efficiently as possible. Roman roads featured a cambered profile with drainage channels at the edges: given that some of these roads are still in existence, it is probably safe to assume that those road-builders of 2,000 years ago had some idea of what they were doing.

Being submerged or permanently wet may adversely affect the structure of the pavement. Depending on what materials have been used, the prolonged presence of water may accelerate the deterioration of the paving. For those materials bedded on sands or grits, inundation and the saturation of the bedding can cause settlement and subsidence,

and drastically reduce the serviceable lifespan of the paving. For hundreds of years, one of the guiding principles of road and pavement construction has been to get the water off and away from the paved area as quickly as possible, and this tenet is as valid today as it has ever been.

## How Much Fall?

Two key questions arise from the need to dispose of surface water: how much slope or 'fall' is needed to drain the paving, and what can be done with the water once it is off the paving?

As ever, the types of material used affect the amount of fall that is necessary. Smoother materials drain more freely than riven or coarsely-textured ones. A pavement with recessed joints, such as setts, cobbles and some types of block paving, tends to retain surface water for longer than a surface such as concrete, bitmac and large flags.

The amount of fall can be expressed in a number of ways. The two most common are as a ratio and as a percentage.

### Calculating Gradients

Ratios are usually written as 1:40 or 1:100 and read as '1 in 40' or '1 in 100'. These figures refer to two components: the 'rise' and the 'run'. Rise is a measure of vertical height (or depth), while run is a measure of horizontal distance. So, a gradient of, for example, 1:40 indicates that there is a rise or fall of 1 unit for every 40 units of distance. That might be 1inch in 40inches, 1cm in 40cm or 1 mile in 40 miles. As long as the units are the same for both rise and run, the

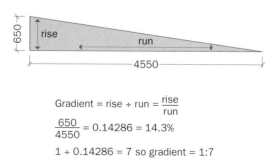

$$\text{Gradient} = \text{rise} \div \text{run} = \frac{\text{rise}}{\text{run}}$$

$$\frac{650}{4550} = 0.14286 = 14.3\%$$

$$1 \div 0.14286 = 7 \text{ so gradient} = 1:7$$

*Gradient ratios and percentages.*

gradient remains the same, and the ratio is dimensionless. The concept can be expressed in a simple equation:

$$\text{gradient} = \text{rise} \div \text{run}$$

Ratios are dimensionless; that is, they need no units to be stated. For some engineering tasks, and for setting-out purposes on site, a percentage figure is preferred to a ratio. A fall or gradient expressed as a ratio of 1:40 is exactly the same as gradient of 2.5 per cent. This is less intuitive; it is possible to visualise a gradient of 1 in 40, but how does one visualise a fall of 2.5 per cent? An example or two should make things clearer.

A percentage gradient is calculated from the same equation stated above. The rise is divided by the run and then multiplied by 100 to convert it to a percentage:

$$1{:}40 = 1 \div 40 = 0.025;$$
$$0.025 \times 100 = 2.5\%$$

So, a gradient of 1:40 is the same as 2.5 per cent.

Another case:

$$1{:}100 = 1 \div 100 = 0.01;$$
$$0.01 \times 100 = 1\%$$

Let us try it with a less convenient number:

$$1{:}63 = 1 \div 63 = 0.0159;$$
$$0.0159 \times 100 = 1.59\%$$

The real advantage of using percentages comes when on site.

**Example 1** On a new patio it is decided that the paving will slope to direct water away from the patio doors. A new gully is to be installed 7.25m from the centre of the doors. How much fall is required?

It is decided that the patio is to have a fall of 1:60:

$$1{:}60 = 1 \div 60 = 0.0167 = 1.67\%$$
$$\text{Run} = 7250\text{mm} \times 0.167 = 121\text{mm}.$$

Thus the top of the gully needs to be 121mm lower than the paving level at the patio doors. Note how the percentage is used in its raw form (0.167). This makes for a simpler calculation, but, if a calculator with a percentage function is available, the percentage value maybe used with the same result:

$$7250\text{mm} \times 1.67\% = 121\text{mm}.$$

**Example 2** A site survey has revealed an old gully on the other side of the patio doors, 8.65m from the centre line. The top of this gully is 150mm lower than the paving level at the patio doors. Will it be possible to lay the paving with adequate fall to this gully?

$$\begin{aligned}\text{Gradient} &= \text{rise} \div \text{run} \\ &= 150 \div 8650\text{mm} \\ &= 0.0173 \\ &= 1.73\%\end{aligned}$$

This is slightly greater than the required fall of 1.67 per cent, and so it will be possible to utilize the old gully.

*Minimum Falls*

The British Standards for pavement construction give minimum 'gradients' to be used for both longitudinal fall (end to end) and for transverse fall (side-to-side). For most pavements the figures are:

| | |
|---|---|
| longitudinal fall (endfall) | 1:80 or 1.25% |
| transverse fall (crossfall) | 1:40 or 2.5% |
| | (*source:* BS7533) |

However, these figures have been determined for use on commercial pavements, where the flow of surface water can be expected to be more channelized and the degree of laying accuracy may be lower than that required on a private project. This may sound contrary, but consider: on a public footpath, no-one is going to notice, let alone lose any sleep over a shallow puddle that hangs around for 15–20 minutes or so after the rain stops, but, on a patio or driveway, this is just the sort of annoying imperfection that can incense the property owner faced with observing the defect every day of their life. Consequently,

| Recommended Minimum Falls | |
|---|---|
| **Type of Pavement** | **Minimum Fall** |
| block paving | 1:60 |
| flat flagstones | 1:60 |
| riven or riven-effect flagstones | 1:50 |
| setts, cubes and cobbles | 1:50 |
| hand-laid bitmac | 1:50 |
| machine-laid bitmac | 1:75 |
| float-finished concrete | 1:75 |
| pattern-imprinted concrete | 1:60 |
| resin-based paving | 1:75 |

it is recommended that the minimum values for gradients on public and commercial footpaths, given above, should be increased slightly, to ensure perfect drainage.

After many years of installing thousands of patios and driveways throughout Britain and Ireland, the following table gives minimum fall values that have been found to give good results on most projects. It does not really matter whether it is longitudinal fall (endfall) or transverse fall (crossfall) as long as there is some fall.

## SPECIAL FEATURES

Finally in this section on design considerations, it is time to look at those 'incidental' items that are often associated with, or incorporated into, a typical residential patio or driveway.

Just where the line is to be drawn between paving and other items of hard-landscaping is debatable. Walls, for example, may often be included in a hard-landscaping project, but their use, design and construction fall outside the scope of this book. For more information, consult *Brickwork and Paving for House and Garden* by Michael Hammett, Crowood Press. Similarly, pergolas, fences and retainers, for example, are not essential components of paving and hard-landscaping projects, but certain structures and fittings ought to be included in any examination of paving construction and design.

### Steps and Terraces

Steps are often an essential component of a paving job. Their most usual function is to link a higher and a lower section of a pavement, but they also be used when not strictly necessary to create design interest in larger expanses of paving by breaking it up into smaller areas at varying levels, an effect known as terracing.

*Steps and terraces create interest and provide a different perspective on the garden. (Stonemarket)*

Consider a proposed patio that would have a gradient of around 1:20. This is a steady slope, less than the maximum recommended gradient for wheelchair ramps (1:12) but a little excessive for use with a patio table, since wine and other liquids tend to splash over the sides of a glass. Such a gradient would be acceptable, but, if space permits, to reduce the gradient to a more wine-glass friendly 1:40 and introduce a number of steps would enliven the design.

Terracing – the use of single steps to move between levels – is a remarkably effective method of injecting interest into what could otherwise be a plain and boring canvas of paving. Even on flat sites, terraces can be built into a patio to subdivide it into smaller areas offering elevated views and varying panoramas of the overall design. Terraces work best when single steps separate broad 'platforms' of a metre or more. Broad platforms between successive steps give users an opportunity to steady themselves and reassess the site after ascending or descending the step. They are non-strenuous, gentle alterations in perspective that emphasize the third dimension of the total landscape.

## Designing

Steps need to be designed in as much, if not more, detail as the rest of the paving. They need to be structurally sound, aesthetically pleasing, but, above all

else, they need to be safe. A major factor in the safety of steps, regardless of whether they are a complex flight or a single step, is the height of the step. Too great a height makes the use of the step uncomfortable and possibly treacherous, but too little has the effect of creating a trip, rather than a step.

To understand how steps are designed and constructed, some familiarity with the terms used when describing them is essential. Of paramount importance are treads and risers. The treads are the platforms on which users stand, while the risers are the vertical lifts between successive treads. Treads need to be of a depth, or 'going', that offers safety and support, while risers need to be manageable for the majority of users. A good, generalized guide would be that each tread should be capable of accommodating a man's foot, so that it is possible to stand on the tread without overhanging the leading edge, which can invoke a sense of unsteadiness. This is generally taken to be 300mm (12in), although shallower steps are used in some circumstances. With regard to the risers, it is assumed that the most comfortable 'lift' for the majority of adults is between 150mm (6in) and 225mm (9in). Although it is possible to reduce the lower end of that size range to as little as 75mm (3in), anything less risks being a trip, while a riser in excess of 225mm becomes challenging and demands too much exertion from the users.

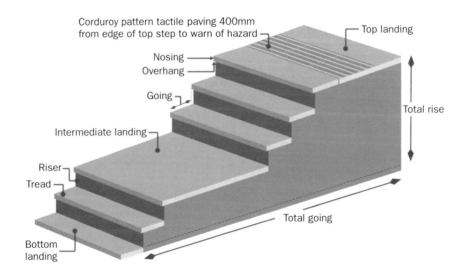

*Terms and definitions for steps.*

There is usually a landing at the top and one at the bottom of a flight of steps, and on extended flights there should be intermediate landings every 1200mm of rise.

### Document M: Disabled Access

It should be pointed out that steps constructed as part of the main or primary access to a dwelling ought to comply with the requirements of Approved Document M: Access to and Use of Buildings, which forms part of the Building Regulations for England and Wales. This document imposes specific design limitations on the use of steps and ramps, the most pertinent of which is that steps should have risers of between 75 and 150mm, and treads with a going of not less than 280mm. Where steps 'spiral' or taper, the going is measured at a point 270mm in from the inside face of the step. Document M also requires a tactile warning at the top of any flight of steps and that there be a contrast in brightness at the nosing of a step that helps visual identification. While Document M may not always be applicable to steps built in the garden of a private home, the advice given is sound, and provides good guidance for the construction of steps for able-bodied as well as disabled users.

The remainder of this section assumes that the steps being designed or constructed are not intended to provide primary access to a dwelling. This allows the design to have risers of more than 150mm and means that we need not include the tactile warning pavers.

### Calculating the Requirements

Whenever more than a single step is used, the key to safety and security is regularity: Whatever sizes of tread and riser are used, they should be uniform and regular. If a 'lift' of 300mm is required, this is best achieved as two steps of 150mm, rather than, say, one step of 125mm and then a second of 175mm. The more steps involved in the flight, the more important uniformity becomes because our bodies are programmed to expect the next step at roughly the same lift as the preceding one, for there to be a rhythm to the steps. It is when steps are irregular, or arrhythmic, that accidents become more likely.

The designing of steps involves two concepts

*Steps need to be stylish, functional and safe. (Acheson-Glover)*

introduced in the previous section regarding gradient, namely, rise and run. However, when describing steps we talk about 'the total going', rather than 'the run', but the concept is the same. To determine how many steps are required for any change in level, it is essential to know the level difference and then to divide this by the proposed height of the risers. As mentioned previously, risers may be of any height between 100mm and 225mm, with 190mm being the ideal for able-bodied users. When working with a blank canvas it may be possible to construct steps that have risers of exactly 190mm, but for many projects steps are required to link predetermined levels and so some design tinkering may be required to create a best fit.

**Example 1** A flight of steps is to link a patio with the lower garden. A site survey finds that there is a difference in levels of 1100mm between the two areas.

Given the 'ideal' step with a riser of 190mm:

$$1100mm \div 190mm = 5.8.$$

This reveals that, if the 190mm ideal is used, five full steps would be needed, along with a 'make-up' step equal to approximately 0.8 of a full step:

$$0.8 \times 190mm = 152mm,$$

which is almost 40mm less than a standard step.

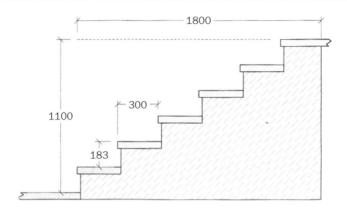

*Step design for Example 1.*

However, if we accept that six steps with equal risers will be used, then:

$$1100mm \div 6 = 183mm,$$

which is not too far from the ideal of 190mm.

While this tells us what size of riser will be used, it does not determine the going of the treads, and therefore the overall size of the flight of steps. If we assume a minimum going of 300mm, then six steps give us:

$$6 \times 300mm = 1800mm.$$

On some projects, the separation between two areas that are to be linked by a flight of steps is fixed, and so the depth of tread can be calculated as part of the design process.

**Example 2**   On a property built on to a steeply sloping site, it is proposed to subdivide the rear garden into a series of terraces separated by retainer walls and linked by steps. The uppermost terrace will be 1200mm higher than the lowest, and there is up to 3 linear metres of space available for the going. Using the ideal riser of 190mm, we can calculate the approximate number of steps required as:

$$1200 \div 190 = 6.3.$$

This suggests that six steps should be used and each

will have a riser of approximately 200mm, just 10mm more than the ideal. If six steps are to be used and there is 3m of space in which to build them, then the tread length is determined by dividing the space available by the number of steps:

$$3000mm \div 6 = 500mm;$$

each step should have a tread of 500mm.

We have now seen how riser height and tread depth can be calculated, but one other factor needs to be taken into consideration: each tread needs to be constructed with some fall to ensure that any surface water does not linger on the tread, where it could be dangerous. By constructing the treads with endfall we can be sure they will drain properly and be safer to use in wet or wintry conditions.

**Example 3**   A flight of steps is to be constructed with treads that will each have a going of 600mm. The total height of the steps will be 850mm. Starting once again with the riser calculation:

$$850 \div 190 = 4.5.$$

This is in the middle – do we use four steps at 212mm (including fall) or five at 170mm (including fall)?

$$\text{Riser height} + \text{fall} = \text{step height};$$

given that each tread will be 600mm in length, if we

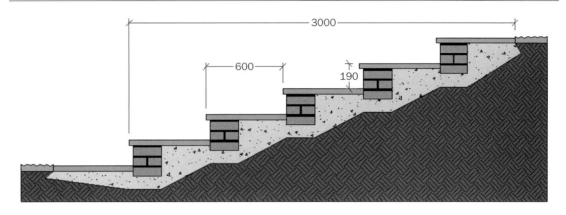

*Step design for Example 2.*

use a minimum fall of 1:40 (2.5 per cent or 0.025), then each tread should have an in-built fall of:

$$600 ¥ 0.025 = 15mm.$$

If that 15mm of fall is subtracted from the step height previously calculated, then we could use four steps at

$$212 - 15 = 197mm,$$

or five at

$$170 - 15 = 155mm.$$

The option to use four steps gives a riser height closer to the ideal of 190mm, and so it is decided to use this format. This calculation has suggested that each riser should be 197mm in height, but, in reality, each step could be constructed with a riser of 190mm and then make-up the remaining 22mm as endfall along the tread, which is equivalent to a fall of:

$$22 ∏600 = 0.0375$$
$$= 3.75\%$$
$$= 1:26.7.$$

The design of steps often involves adjustments to the 'ideals' to come up with a design that fits the site and fulfils as many of the 'ideal' requirements as is possible.

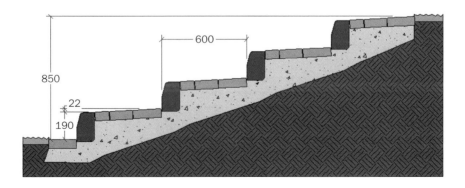

*Step design for Example 3 – incorporating fall.*

*Handrails*

It should be noted that, when flights of steps are to be constructed, and especially those involving three or more steps, then a handrail is strongly recommended, and, to be fully compliant with Document M, a handrail is mandatory for all flights of steps consisting of three or more steps. Obviously, the type and layout of the steps will determine what form of handrail is best suited, but it needs to be fitted to at least one side of the flight, at a height of 900mm above the 'pitch line' of the steps and extend by a minimum of 300mm beyond the top and the bottom step.

*Tread Width (Going)*

So far, all that has been considered is the cross-sectional detail of a step or flight of steps, which involves the riser height and the tread depth, but the width of the steps also needs to be considered. As a general guide, the minimum width of a step should be sufficient to allow the safe use of the structure: an absolute minimum width of 600mm is recommended for garden and patio steps (Document M requires 1000mm), anything less feels narrow and constricted. There is no maximum width – the steps may be as wide as is needed to suit the site or the styling of the design; but with flights of steps, the comments above regarding the provision of handrails should be borne in mind, and, with wide flights, it may become necessary to install intermediate handrails for reasons of safety.

## Ramps

Not everyone can cope with steps, and it is now part of the Building Regulations for England and Wales that 'reasonable access' should be provided to dwellings and public buildings. The simplest and most straightforward way of implementing this is to use ramps in place of steps, wherever possible. Ramps are basic structures, being nothing more than a sloping pavement. The requirements set out in Document M, while not strictly applicable to steps and ramps used in gardens and private driveways, provide good, general guidelines for the construction of ramps that are suitable for both the able-bodied and those less agile. There are a number of key requirements stipulated in Document M, namely, ramp width, ramp gradient and the provision of landings. There is also a statement making it clear that 'gravel and loose-laid shingle' are not suitable for ramps. Hard, bound or firm paving should be used in all instances.

With regard to width, Document M requires a minimum of 1000mm, although many ramps are wider to ensure easy access for wheelchair users and their companions. Ramp gradients should not be steeper than 1:12 (8.3 per cent) and preferably between 1:20 (5 per cent) and 1:15 (6.7 per cent). Ramps must have a landing top and bottom, which must be level, have crossfall not exceeding 1:40 (2.5 per cent) and be at least 1200mm in length. Intermediate landings are required where the length

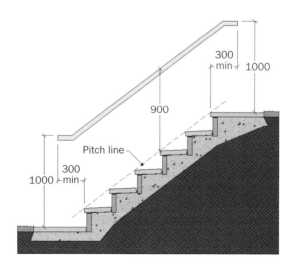

*Handrail requirements.*

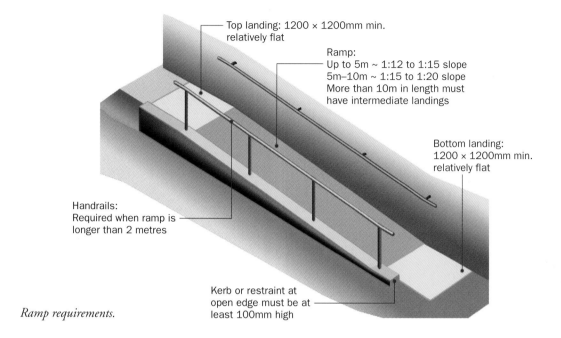

Top landing: 1200 × 1200mm min.
relatively flat

Ramp:
Up to 5m ~ 1:12 to 1:15 slope
5m–10m ~ 1:15 to 1:20 slope
More than 10m in length must
have intermediate landings

Bottom landing:
1200 × 1200mm min.
relatively flat

Handrails:
Required when ramp is
longer than 2 metres

Kerb or restraint at
open edge must be at
least 100mm high

*Ramp requirements.*

of the ramp exceeds 10m for gradients up to 1:15, and at 5m intervals for those ramps having a gradient between 1:15 and 1:12. Again, these landings should be at least 1200mm in length, with crossfall not exceeding 1:40.

For larger or longer ramps, two safety items may need to be considered: the provision of a handrail can be invaluable for those termed 'ambulant-disabled', generally meaning those reliant on using a stick or crutches. Handrails make it easier to haul oneself up the ramp or ease oneself down, and, if a handrail is included, then some form of kerb would be a useful addition, since its presence ensures that the wheels of wheelchairs do not run into the stanchions. Raised kerbs are useful even if a handrail is not provided since they can help to ensure that wheelchairs, children's cycles, and the children themselves, do not go tumbling over the raised edge.

## MANHOLE COVERS AND OTHER ACCESS FITTINGS

As mentioned previously, drainage is the key to a successful pavement, and it is difficult to construct a pavement of any size without incorporating one or more 'drainage fittings', the collective term for manhole covers, grids, gratings and all the other elements used to remove surface water from the paving.

However, depending on the type of paving that is to be used, the drainage fittings may sometimes appear unsympathetic, or may even disrupt the design and layout. The clearest examples of this are the concrete inspection chamber covers commonly found on properties built since the 1950s. While these covers may look just about acceptable when laid within an area of plain concrete flags, they look truly awful when left unchanged in

*Concrete manhole covers are rarely attractive and have the unfortunate habit of being as discreet as the proverbial sore thumb.*

*Recess tray covers can make manholes and access chambers almost 'disappear' within the paving. (ClarkDrain)*

*Modern gully fittings tend to be much smaller and neater.*

a more decorative surfacing, such as block paving, and, within an area of riven flagstones or setts, they not only look totally out of place in terms of period and materials, they can appear as positively ugly.

### Recess Covers

Thankfully, the use of recess tray covers has now become the norm. These are covers that consist of an outer frame inside which sits a tray that is filled with whatever type of paving is used elsewhere on the pavement. The result is a much more discreet fitting, that is not totally invisible but at least does not distract one's attention, as is often the case with the older, concrete fittings. The price of these recess tray covers has become much more reasonable over recent years as their usage has increased. In the 1980s, even a simple 600 × 450mm tray could cost in excess of £50, but the present-day version now retails for nearer £25, and will be generally engineered to much better standards. They are also available in a huge range of sizes, and it now makes economic sense to use a 300 × 300mm tray to replace even the smallest standard inspection chamber cover.

Of course, ductile or cast iron covers are still used and look perfectly acceptable with some surfaces, such as bitmac or stone, and even steel covers can look satisfactory within a complementary surface, despite the shiny, galvanized finish.

### Gratings and Gullies

Other common drainage fittings are gratings and gullies, which are referred to as 'grids' in some parts. The gully is the sub-surface chamber that holds the water and directs it into the drainage system, while the grating (or grid) covers the hole into which water drains and keeps out leaves and litter. The gully may have a hopper fitted to it, which may be thought of as a funnel that collects water from a wider area. Some of the older hoppers found on pre-war properties are quite attractive and may be reused, provided that they are undamaged and serviceable, but any design process should consider whether the use of new gullies, hoppers and gratings could give a better standard of finish.

### Linear Channels

Next in this brief survey of drainage fittings, linear channel drains deserve a mention. Like recess trays, these too have benefited from cost reductions brought about by wider use, and versions suitable for residential driveways can now be bought for less than £10 per metre. Some thought is needed when the possible use of a linear channel drain is suggested: laid against an edge or a wall, or across a garage threshold, they can look neat and discreet, but, if stuck in the middle of a driveway or a patio, they can disrupt a design and look unsightly. Where there is no viable alternative other than to use a linear channel drain in such a position, one of the 'slot' varieties may be a better choice.

When using one of the U-shaped channels supplied with a grating, pay particular attention to the grating itself: the channel component is buried and can hardly be seen when it is installed, but the grating is on show and so needs to complement or at least blend with the rest of the paving. There are two main types of grating for use with these channels: galvanized steel or plastic/composite. Which looks the better is a matter of personal taste, but many feel that the black plastic/composite gratings are more discreet than the shiny, galvanized versions. However, there are different colours available, both in the plastic/composite forms and as powder-coated steel, although these generally cost more than the standard units. It is also worthwhile assessing the general quality of the grating. Some of the cheaper, 'budget' offerings save on production costs by offering thin or 'tinny' steel gratings that seem to flex and deform as soon as a lawnmower runs across them. Any gratings used on channels that will be trafficked by the family's vehicles should be at least Class B products and marked as such.

## Dished Channels

Finally, a brief mention for dished channels. These are typically pre-cast concrete or fired clay units featuring a shallow, concave, 'dished' profile that collect and direct surface water to a convenient disposal point. Uncoloured, 'natural' concrete units are often used on commercial schemes, but, for residential driveways and patios, the coloured concrete or clay paver units are much more attractive and provide a useful solution at low cost, and with simple installation.

## PLANNING AND DRAWING

Having looked at some of the less obvious details that ought to be considered when planning a path, patio or driveway, we come now to the actual planning and design process. Simply pacing out the area, guesstimating the quantity of materials and forging ahead with the work is unlikely to produce the best results. Some pre-planning and design are essential, but there are a number of different ways in which this can be done.

### Site Survey

Whatever design method is selected, it will require a site survey as the first step. This involves measuring and plotting out the site, taking note of any pertinent features, and transferring all this information into a drawing. On more complex sites it may be necessary to determine site levels, as well as dimensions, so that falls, gradients, steps and ramps can be planned as part of the design process.

*A linear channel is connected to the main drainage system by means of a trap fitting.*

*A dished channel laid against a kerb edging forms a neat and effective drainage option. (Marshalls)*

Most residential projects can be planned in just two dimensions: in plan view, also known as the bird's-eye view. All that is required is a tape measure, a notepad, a pencil and a preferably glamorous assistant, known as a 'chainman'. The tape measure needs to be sufficiently long enough to allow the dimensions to be taken as one single measurement, rather than by struggling with, say, a 3m tape from the local DIY shop and then adding together a series of measurements to establish the length of the side of the house, or the distance between the back door and the greenhouse. For the majority of patio and driveways, a 30m tape will be adequate.

Start by sketching the site. Using graph or quadrille paper makes the task much easier. Do not worry too much about drawing accuracy: as long as the layout is approximately correct, it can be tidied up once the measuring is complete. First, record the fixed features and their relationship to each other. This will normally include the house, the garage (if present), and any other relevant buildings, such as sheds, stables, outhouses and greenhouses. Also mark boundary walls, fences, hedges as well as any flowerbeds, lawns or garden features that may affect the design.

Once the area has been roughly sketched out, the measuring may begin. Again, do not get bogged down with accuracy; measure to the nearest 25 or 50mm, rounding off as necessary. If the front of the house is measured and found to be 7890mm long, then recording this as 7.9m will be satisfactory. Measure all the fixed features and record the measurements on the sketch. Have the chainman hold one end of the tape while the other is paid out to the measuring point and record each measurement immediately. Trying to to retain a figure in memory while a second or third measure is taken is almost guaranteed to result in errors. Measure and mark the position of important features, such as doorways and gateways in walls, or the position of existing gullies and drainage/access covers.

Once all the fixed features have been recorded, the distances between them should be measured. Take as many measurements as seems reasonable, since they act as a check on previous measurements when plotted out. Use triangulation to define specific points by measuring the distance from the point to at least two other, known points. Break up non-rectangular areas into triangles and measure their axes. For fast arcs, try to determine the radius; slower arcs and curves are better noted as a series of points at, say, 2m intervals, and then recorded by triangulation.

## Preparation of a Drawing

Once all the measuring has been completed, a new, 'working' drawing can be prepared by taking the measurements recorded on site, scaling them as appropriate, and then plotting them out with pencil and paper (along with rulers, compass and eraser), or by using a computerized (CAD) system. There are pros and cons to whichever method is used, and so there is no definitive 'best way'. Multiple measurements can act as checks to ensure that the figures recorded actually tally when plotted, but allow minor errors to pass without worrying too much. Regardless of how accurately the measuring and plotting out is done, it is an absolute certainty that the completed paving will not correspond exactly – there is bound to be some error and some deviation, so there is no point in being worried about it.

For the home DIY-er planning their one and only ever patio or driveway, to invest in CAD software and then spend hours (if not weeks) learning how to use it, may not be the most appropriate option, but, for a contractor or professional designer, there is no doubt that CAD can and does speed up the design process (once you can remember where all the tools are located and which button does what), and the completed drawing is often more professional in appearance, as well as being much simpler to modify if and when additions or amendments are required.

A third option that often appeals to DIY-ers is to use one of the 'landscape designer' software packages that are available. These range from free trial packages found on the CD-roms given away with magazines, up to highly complex, dedicated CAD-style programs. There are packages available for under £50 and there are 'paving designers' given away by some of the bigger manufacturers.

Generally speaking, you get what you pay for: the free-trial products will probably have some of the features disabled or will have the print function missing. The celebrity-endorsed design packages are

## Triangulation

Measuring and recording the length of walls that are 'square' or perpendicular is a simple task, but many projects will involve walls at odd angles, or will need to take account of the position of some item that is not conveniently located against a wall or an edge. Situations such as this call for the technique known as triangulation. This relies on creating simple triangles with the object to be recorded at one vertex of the triangle, while the other two vertices are located at known and fixed points elsewhere on the site. In the example shown, the wall on the left-hand side 'kinks' out at an unknown angle. The length of the wall, labelled A–B, can be measured, but this is not enough

information to enable the angle to be plotted. However, by creating a simple triangle, A–C–D, the position of A can be recorded relative to the fixed points at C and D (the internal corners of the walls). In practice, a third 'check' measure would probably be made, from A to E, to ensure a greater degree of accuracy when plotting out the site. Similarly with the inspection chamber: its position can be recorded by making it one vertex of a triangle C–F–D. Again, an additional check measure could be taken to ensure accuracy; E–F or A–F would provide a check on the measures taken for C–F and D–F.

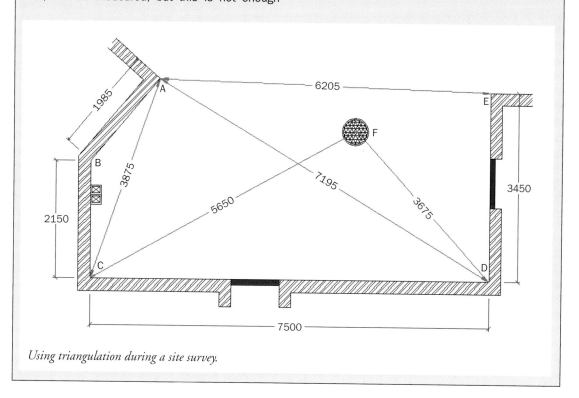

*Using triangulation during a site survey.*

often thrown in the bin in disgust, as the lavish claims made on the packaging rarely match the actual output, even after a full weekend spent learning how to use the product. The manufacturer-supplied give-aways are typically limited to the products of that company and tend to repeat the same pattern or design over and over again. The fully-functional,

all-singing-all-dancing landscape designer suites can be as complicated as a CAD program, or even more so.

The best way is to use what is at hand, what makes you feel comfortable, and that which allows you to express your creativity as simply as possible. If this means pencil and paper, then that is fine. If the drawing produced is intended only for your own use, what does it matter if it is not a work of art? As long as you understand it and it has made you think through the actual design and construction, then it has done all that is required.

## Using Professionals

However, if all this sounds too daunting and your head is spinning already, then perhaps to call in a professional designer or a contractor offering a basic 'design and build' service may be the answer. A professional designer offers the possibility of having the design work done for you, and then turning over the actual construction to DIY, with the benefit of having a professional eye look over the job and think through the design and construction processes that will be involved to ensure that what is planned is actually feasible. Professionals bring a greater awareness of materials, of what works and what does not, of how best to drain an area and what to use in so doing. They also have a knack of spotting potential problems that may easily be overlooked by the DIY-er. How will the edges work against that fence? If the rear patio is levelled out, will the gates still open properly? What is the best position for that new gully that is needed on the driveway? These are the sorts of thing that a professional designer can identify and resolve before the ground is opened, and it can often be money well spent.

Similarly, many of the better specialist contractors offer a design and build service. Just as there is a vast range of skill levels among contractors, so there is a vast range of design skills. Some contractors have the habit of 'designing' the same drive or patio time and time again. They choose materials that suit them or which offer a better profit margin, rather than those that are actually the best suited to the job in hand. However, there are also some highly skilled contractors that offer an in-house design service that involves detailed discussion with the client, covering most, if not all, of the topics aired in this chapter, and, over a period of time, a design should emerge that delights the client and is eminently buildable by the contractor, who by this stage is intimately familiar with the design.

Naturally, there is no such thing as a 'free' design service. When a contractor is asked to develop or propose a design, the cost will be built into the total project price. Should the actual construction work then be awarded to a different contractor, it is only right and fair that the original designer-contractor should be compensated for the time and effort spent in developing the design.

Also bear in mind that a design and build contractor may 'persuade' a client to use a particular type of paving, perhaps one that offers a better profit margin, or is more convenient to the contractor. For this reason, it is always best to explore what is available, at what price, and ensure that the contractor explains just why such-and-such a product is better or more suitable than the other options. There is also a tendency among contractors to denigrate certain products with which they dislike working. Maybe the client fancies clay pavers, but the contractor knows how awkward they can be to cut and align accurately, and so may try to dissuade the client and steer them towards the easy option – concrete blocks.

Even the best and most impartial of designers and contractors have favoured products. Previous work will have shown them that certain products or materials work best with particular types of property, and, although the advice they offer is generally sound, it should not be regarded as definitive. It is the client who has to live with the paving, seeing it every day for the next umpteen years and so it is more important that the client receives what they want, which is not always the same as what is wanted by the designer or the contractor.

In summary, when it comes to preparing the design, the typical property owner needs to assess his or her own skills and weigh these against the cost of bringing in professional help. A couple of hundred pounds or so spent on having a site surveyed and a basic design submitted may be a sound investment if it saves the bother of being out there in the wind and rain, measuring up, and then wasting a weekend trying to make sense of the scrawled figures and partial notes. A basic layout design can be copied and overlaid with your own design: the hard work of measuring and plotting out will have been done – sketching in ideas, patterns, possible designs and potential features can then become an enjoyable challenge instead of a stress-inducing, blood-pressure-raising nightmare.

# CHAPTER 4

# DIY? Contractor? Costs?

## CHOOSING BETWEEN DIY AND A CONTRACTOR

This chapter considers the pros and cons of using a contractor or opting for DIY. The DIY section introduces the skills and tools required, and the possibility of DIY-ing some or all of the work is considered.

The section on the contractor option looks at how to select a shortlist of potential contractors, how to ensure that the contractor fully understands the design by using contract documents to formalize the relationship, how to be certain that the bid-submitting contractors are all pricing the same job, and then how to assess the bids. For the professional contractor, this section outlines the reasonable expectations of the general public and the value of professional presentation.

Earlier, the arguments for and against DIY design or bringing in the professionals were considered. This chapter takes that one step further to consider whether DIY or a contractor should be used for the actual construction work.

It is not a straightforward either/or decision: there are a number of factors needing to be thought through. For those not given to DIY, employing a contractor might seem the obvious choice, but some projects may be considered too small to warrant the attention of a professional. Many contractors are not interested in getting involved in projects of less than 30m². The amount of planning and paperwork involved in such a project, measured against the potential profit, renders them unattractive to the more professional companies, and so they tend to be left to the odd-job or handyman type of builder.

Even if leaving the work to a contractor would be the first choice, it may not actually be possible to get one to take it on. At the other end of the scale, there are projects that are far too large to be tackled as a DIY enterprise.

Difficulty experienced in finding a reputable contractor, or budgetary constraints, might make it seem that the only way to ever get the paving or landscaping completed is to rely on DIY skills, and hope that friends and relatives will pitch in to help. Knowing the limitations of what is possible and what is not can help to ensure the right decision is taken.

## THE DIY OPTION

### Fitness and Competence

Obviously, one of the first considerations is the fitness and skill level of the candidate DIY-er. Paving, drainage, hard-landscaping and the range of skills that come under the banner of 'groundworks' are among the most physically strenuous and demanding jobs in the building trade. A typical groundworker is said to have a daily calorie demand higher than that of a face-working miner, and, even with modern machinery, diggers, vacuum lifts and power saws, it is still exceptionally hard work. The upside of all this is that it is a great way to keep fit, and cheaper than joining a gym.

Constructing a hard-landscape, even a small patio, involves a lot of digging, lifting, fetching and carrying. If physical work is not your normal routine, you need to consider your level of personal fitness. How confident are you about taking on this work? If you are reasonably fit and active, then building your own patio or driveway can be immensely rewarding. If you are just looking for a challenge that will probably toughen you up, save you money, and trim a couple of notches off your belt, then this could be it.

You must be honest with yourself: just how fit are you? Do you participate in sport on a regular basis? Do you cycle, swim, climb, jog or walk the country-side at weekends? Do you have a job that keeps you in trim and ticking over? There is no point in deluding yourself – if you are unfit, for whatever reason, taking on a hard-landscaping project is a big commitment, and it is not something that is easy to walk away from if you realize that it is too much, or you lose interest. If you think that it is hard to get a contractor to take on your job from scratch, it is ten times harder to get anyone reliable to complete a quarter-completed DIY disaster. Take it from an ex-contractor: we do not want to know!

## Tools

What about the tools, what in the trade is known as 'the kit'? Do you have or can you get hold of the necessary tools, and could you handle the necessary machinery, the 'plant', without demolishing the house or killing the cat? Although some of the more straightforward paving projects could be carried out using the tools found in most tool sheds, some paving types, such as imprinted concrete or tarmac, require specialist kit that may be difficult for the average private person to acquire. The tools needed to complete most paving projects are considered in Chapter 6.

## Timescale

Consider also the timescale of the work and the consequent disruption to leisure time and day-to-day family life. A contractor can complete a 50m² driveway in four or five days. Assume five days, with three men, working for 8 hours per day, this gives a total of 120 manhours. One person working with an assistant, both sacrificing their weekends to work on the project, would need to give up four consecutive weekends to achieve as much, and that is assuming a level of productivity matching that of a professional contractor. In reality, the productivity of a competent DIY-er is around half that of a tradesman, so the four

weeks estimate is likely to be nearer eight – that is two months of sacrificed weekends. Is it reasonable to expect DIY-ers to give up all their free time for so long a period? And what if the weather is less than favourable? Two months may easily become three if there are a few wet weekends or the work is being done over the winter.

## Family

It is not just the commitment of the workers, the whole household has to endure weeks of trudging across mud, crushed stone, sand and concrete, as well as having no place to park the cars, and just to get anything out of the garage can be a nightmare.

How much is your time worth? This is not the same as the hourly rate you can earn, but the value you attach to your free time. Calculate the labour charge incurred should a contractor be used, and divide this by the number of hours you would need to put in to complete the job: now how much is your time worth? Of course, the timescale can be reduced by putting in extra hours during the evening, assuming that there is sufficient light, and that a day at the paid job has not drained away any desire for additional effort. However, if you consider that it is easy to spend 30 minutes just getting out and putting away the tools needed for each shift, the potential productivity of short stints can be heavily affected. Bear in mind also that enthusiasm dwindles, and after two or three weeks of sacrificed weekends and extended evenings, the thought of going out there again when there is a good film on at the cinema, or there is a pint with your name on it at the pub, does not seem all that attractive.

## Personal Achievement

Against all that should be weighed the enormous sense of achievement that comes from planning, sourcing and building a patio or driveway yourself. The end result is something that should give count-less years of service and satisfaction. For someone unaccustomed to physical labour, pavement construction is an activity that can be carried out at a pace to suit the individual, with not much that can go disastrously wrong. It is a life-affirming challenge for anyone more familiar with pushing a pen or a mouse rather than a wheelbarrow, and even a simple, twelve flagstone patio can engender a tremendous boost to self-confidence – and that is not to mention the money that can be saved.

## THE CONTRACTOR OPTION

The greatest concern regarding contractors in any of the home improvement trades is to ensure the one selected is competent, diligent, honest and reliable. We have all heard horror stories about cowboy contractors, but the truth is that the great majority of tradesmen do a fair job for a fair price, it is just that tales about workers doing the job on budget and on time are not as newsworthy as the stories of woe and disappointment that colour the popular image of the building trade.

### Finding a Contractor

So, how does one find that wonderful contractor? The first thing to consider is whether to use a general landscape contractor or a paving specialist. General contractors can usually turn their hand to many different types of paving and surfacing, as well as brickwork, water features, pergolas and fences, turfing and planting, while the specialist, as the name implies, concentrates on one or two specific aspects of the hard-landscaping trade, such as block paving or sett laying.

Does the work in hand involve, solely or predominantly, one or two types of paving, or is it a mixed scheme, involving hard and soft landscaping? The answer to this fundamental question can help to determine whether a pavior or a landscaper is required. The two trades do overlap, but they tend to advertise in different places and affiliate themselves to different industries. Most paviors regard themselves as part of the construction industry, while many landscapers feel that they belong to the horticultural trade. As a general guideline, driveways are nearly always a job for a paving contractor, patios are undertaken by both paviors and landscapers, while garden make-overs are usually best suited to the landscapers. This basic guidance can help you decide just which sector of the trade is most likely to be interested in your project.

Many paving contractors describe themselves as 'block paving specialists'. This is undoubtedly true in most cases, but there are those for whom block paving is their *only* skill, not necessarily one in which, after due reflection and consideration, they have elected to concentrate their alleged talents. Similarly, some landscapers need to be reminded that it is 'green side up' when it comes to laying turf, yet they manage to convince themselves that the ability to plant shrubs makes them experts. There are the good, the bad and the indifferent in both trades, so caution is required.

### The Shortlist

Having decided whether to look for a generalist or a specialist, the next task is to draw up a shortlist of candidates before inviting them to survey the site and submit a price. There are a number of sources, but the best possible recommendation is, and always has been, word of mouth. Talk to friends and family to ascertain whether anyone has had similar work done, and whether they can suggest any contractors to approach, or to avoid. Maybe someone at work knows someone, or could you put up a request for information on the canteen noticeboard? Have you seen a driveway or landscaping job locally that has taken your eye? Ask the property owner for his opinion of the contractor, good or bad.

Have you noticed a local company that has been on the scene for at least a couple of years? Many contractors use signboards to promote their presence on jobs and these are remarkably efficient in generating 'brand awareness' in local areas, particularly when the same company name has been popping up in the locality for a period of time. Again, stop and ask present or previous clients for their opinion and experiences.

Using a local contractor is usually a good strategy, as they are more likely to want to maintain their reputation in the area than would a regional or national contractor who may be working at the other end of the country next week. Furthermore, should any basic maintenance or remedial work be required, a local contractor is easier to get back on site.

There are 'approved lists' managed by some of the professional bodies and certain manufacturers. These differ in their entry requirements, but, generally speaking, they do tend to sort the wheat from the chaff. Approved lists cannot guarantee a trouble-free and perfect job, but they do reduce the chance of being duped, let down or conned by a rogue. Some lists are available via websites, others by making a telephone call, but they all work on a postcode basis and there is usually some form of vetting involved before contractors are added to the list. The service is usually free to enquirers, being funded by annual membership fees payable by the listed contractors. Some of the lists include an arbitration service, and/or an insurance-backed warranty service, which may be chargeable. Whether it is fair and reasonable to pay an additional fee for a guarantee, over and above the cost of the work, is a matter of some debate.

Other sources for potential contractors include the telephone directory, in the paving and driveway contractors and the landscapers sections. Again, the fees charged to be included in the commercial telephone books often deter those intent on defrauding people, but such entries do not guarantee reliability. Sadly, the few hundred pounds paid for a fancy-looking advertisement in the directory can soon be recovered if you are fleecing people.

## The Site Assessment

The next step is to invite the contractors to visit the site, to talk about the plans, assess and measure the work, and then submit a price. The better contractors often have a portfolio of their previous work, although some use publicity photographs supplied by manufacturers. Ask whether any of the pictures shown portray the contractor's own work, or are shown 'purely to give an impression of the products that may be used'. Some contractors give verbal prices that are effectively worthless as there is no documentary evidence of what was agreed. Insist on a written quotation or estimate.

There is no real difference between quotations and estimates. Quotations will only cover the obvious work, with any additional or unforeseen work charged as extra, while estimates give an indication of what the obvious work will cost, assuming no additional or unforeseen work. The price document, whether it is labelled as a quotation or an estimate, should state exactly what will be done and for what price. It should state the size of the area that will be paved, and it should also quantify any other works, such as the length of any kerblines, the number of

recess tray covers, or the rate for any additional drainage work.

If you have plans, sketched by yourself or by a designer, give copies to the contractors and ask them to base their prices on these. This helps to ensure that each contractor is pricing the same job, and not their own interpretation of what they think is best for you. Plans and drawings are often quantified; that is, they are accompanied by a document known as a Bill of Quantities that lists each of the tasks to be carried out, the materials to be used, the method of construction and the quantities required. These bills of quantity can vary enormously, and are not always necessary, but, like the plans, they do ensure that the same work is being priced and leave little or no leeway for guessing or interpretation by the contractor. They also ensure that the materials you have chosen are documented and this therefore removes the temptation for a contractor to choose products that offer a better profit margin or a lower cost than those that would be used by competitor contractors.

## Contracts and Warranties

Check whether the contractor uses a contract document and, if so, check the terms and conditions – not all contracts are scrupulously fair, and some used by less reputable traders are decidedly one-sided and stacked against the client. However, there are good contracts available; some are offered by sponsors of the approved lists or from trade bodies, as well as the general building contracts used in the construction industry, such as the 'minor works' contract or the 'building contract for a homeowner/occupier', published by JCT. See Useful Links, page 186.

Also check any warranties or guarantees that are offered. These should be written documents that clearly define just what is covered and for what length of time. They should also state whether they are administered by the contractor or backed-up with an independent insurer – if the contractor were to become bankrupt or cease trading for whatever reason, any guarantee that is not backed by an insurance policy immediately becomes worthless.

## Cowboy Spotting

Beware of any contractor who does not have a landline telephone number. Those with only a mobile

| Item | Description | Quantity | Rate £ p | Sum £ p |
|---|---|---|---|---|
| 1.1 | Excavate over site to formation level and cart all spoil to licensed off-site tip. Remove all vegetation and deleterious matter. Depth of excavation not less than 200mm | 10.9 m³ | £ 90.00 /m³ | £ 981.00 |
| 1.2 | Supply and place permeable geotextile membrane Terram 1000 or equivalent | 54.3 m² | £ 1.25 /m² | £ 67.88 |
| 1.3 | Supply, place, level and compact to finished depth of not less the 100mm, DTp 1 granular sub-base material | 54.3 m² | £ 9.00 /m² | £ 488.70 |
| 1.4 | Supply and lay on 100mm deep bed and haunch of ST1 concrete, 125x125x100mm low-rise block-paving kerbs, colour to be red, with 12mm mortared joints. Laid to straight lines | 26 m | £ 23.00 /m | £ 598.00 |
| 1.5 | Supply and lay 200x100x50mm concrete block pavers in charcoal colour as 200mm wide soldeir edge course to entire perimeter. Includes mitre cuts at corners | 33.2 m | £ 6.50 /m | £ 215.80 |
| 1.6 | Supply and lay 200x100x50mm concrete block pavers in buff colour as 100mm wide band course inside charcoal perimeter | 31.6 m | £ 4.50 /m | £ 142.20 |
| 1.7 | Supply, lay and compact to level on 35-50mm Class M sand laying course, 200x100x50mm concrete block pavers in brindled red colour. Pattern to be 45° herringbone. Includes all cuts, compaction and joint sealing. | 40.32 m² | £ 27.00 /m² | £ 1,088.64 |
| 1.8 | Supply and fit to level, 600x450mm recess tray cover to replace existing IC cover. Includes all cuts | 1 nr | £ 90.00 ea | £ 90.00 |
| 1.9 | Clear site of all rubble on completion | 1 nr | FOC ea | £ - |

| | | |
|---|---|---|
| Sub-total | £ | 3,672.22 |
| VAT | £ | 642.64 |
| TOTAL | £ | 4,314.85 |

*Typical bill of quantities compiled for the example driveway considered in this chapter.*

number are often shady – all good contractors have landline numbers that are actually answered by a human. Knowing this, some of the real cowboys have started to use non-geographic numbers (0800 or 0870) to mask their itinerant habits. Look for the name of the proprietor or the partners on any headed notepaper. Limited companies must have a registered company number, and nearly every contractor worth their salt will be VAT-registered and include the VAT registration number on their notepaper. There should also be the full trading address of the contractor that is easily verifiable.

Other cowboy indicators include pressuring or intimidating clients; if this happens to you, call your local trading standards office. Also be cautious of 'special offers' available only if you agree immediately, and never, ever accept any offer from a cold caller. Good contractors never cold call; they have no need to do so.

## Choosing the Lucky Contractor

Once the contractors have submitted prices, the final decision needs to be made, and it is not always a decision based on cost alone. The credentials and the

*Sometimes, only a contractor can achieve the standard of finished required for that special project. (Stone & Style)*

credibility of the contractor must be taken into consideration. Recommendations and references should be followed up; previous works should be visited and checked. If a contractor claims membership of a trade body, ask for details, call the body in question and ask for confirmation.

A good contractor will be approachable, communicative and cooperative. You have to feel comfortable chatting with him or her and feel they are being honest with you, otherwise you will not be able to maintain a relationship where you are the boss, ensuring that you get the paving or the hard landscape you want, and not just whatever it is the contractor might want to foist upon you.

Check about 'lead time', which is the delay between your confirmation of the order and the expected start date. Most good contractors have a lead time of four to twelve weeks; if they can start tomorrow, why are they not busy? The projected start date should be written into the price document as a 'target', but it would be unfair to expect most small contractors to adhere rigidly to the scheduling: the vagaries of the weather, and the fact that other jobs may be extended or delayed, all combine to make accurate predictions impossible. However, any delay of more than a couple of weeks should be a matter of concern, and, purely out of courtesy, the contractor

should keep you regularly updated regarding the expected start date.

## Contractor Benefits

The advantages of using contractors are legion. They bring with them a lifetime of experience and a vanload of the right tools. They have the awareness and understanding to overcome the unexpected, if and when it arises; they know innumerable workarounds; they have a little black book of contacts that can be used to get hold of labour, plant and materials at short notice; they are usually fitter, stronger and more accustomed to hard work, and they have a productivity rate that the average DIY-er can only imagine.

And that is what you are paying for – not just a bunch of callus-handed ne'er-do-wells, all brawn and no brain, but a skilled and dedicated workforce that can complete your project in a fraction of the time and headache tablets that you would need for a DIY project. A typical drive or patio will be completed in a week or two, rather than a month or two, and you can travel home from your own day's labours safe in the knowledge that there is nothing more strenuous to tackle than boiling the kettle and making a mug of tea for them when you get in.

# CHAPTER 5

# Costings

Whether it is to be DIY or a contractor is to be employed, calculating the cost of the works ought to be done by someone before the ground is broken. Contractors use dozens of different ways to decide what to charge; some simply find out what their local competitors are charging and then match it; some will consult a 'pricing guide' that gives pre-calculated rates for hundreds of different tasks in the building trade, while others will break down each job into a series of items and then subdivide these items to account for the three primary components: labour, plant and materials.

How individual contractors price-up work is immaterial at this juncture, but some understanding of how the cost of a job is determined is a useful exercise that can actually help in planning how the work will be tackled, and also in identifying previously overlooked issues. All costing exercises involve a simple calculation of quantity multiplied by rate to give a total amount. It may be cubic metres of excavation to be removed from the site, the number of blocks needed for the edge courses, or the bags of cement required, they all need to be determined by measurement or mathematics, so that the item cost, and eventually the total cost, can be calculated.

## BREAKDOWN TO LABOUR, PLANT AND MATERIALS

Calculating quantities is a skill in itself, and a good quantity surveyor (known as a QS in the trade) can usually 'take off' quantities for an entire project from a simple, two-dimensional drawing. This is achieved by having a reasonable understanding of how the work is carried out, making allowances for wastage, bulking-up, and loss, and having the ability to break down the item into its component parts. For example, consider the task of constructing an edge course for a driveway. Using clay pavers laid on concrete, the task may be broken down as follows:

## Labour

Estimate two operatives – a skilled pavior and a labourer – lay 120m of this edging per day at a cost of £90 + £55 = £145, then

$$£145 ÷ 120m = £1.21 \text{ per metre.}$$

## Plant

Although basic tools will be needed, there is no requirement here for mechanical plant that might incur a cost.

## Materials

Concrete bed – each edge course uses a 200mm wide brick with 100mm of haunching; allowing for 'spread', assume that the total bed width would be 350mm; the bed is to be 100mm deep, therefore, for each linear metre of edge course:

$$\text{length × width × depth}$$
$$= 1m × 0.35m × 0.1m$$
$$= 0.035m^3.$$

Concrete haunch – 100mm wide, finished within 15mm of the top; pavers are 65mm deep, hence haunch must be 50mm deep, therefore the requirement for haunching is:

$$1m \times 0.1m \times 0.05m = 0.005m^3$$

This gives a total requirement for concrete of:

$$0.035m^3 + 0.005m^3 = 0.04m^3 \text{ per linear metre}$$

Concrete costs a rather convenient £90 per m³, therefore:

$$0.04 \times £90 = £3.60 \text{ per linear metre}$$

The bricks used to form the edge course are clay pavers, 200 × 100 × 65mm and bought at a cost of £23 per m²; they are to be laid 200mm wide, giving 5m of edge course from each square metre of pavers:

$$£23 \div 5 = £4.60 \text{ per linear metre}$$

These costs can now be summarized:

| | |
|---|---|
| labour | £1.21 |
| plant | – |
| materials | |
| concrete | £3.60 |
| pavers | £4.60 |
| | £9.41 |

This calculation has been simplified for the sake of clarity. No allowance has been made for any excavation that might be needed, and a price for ready-mixed concrete has been used although for a DIY project, it is quite likely that the concrete would be mixed on-site, but the principle remains correct. The task is broken down into a series of components; the quantities of each component is calculated and then costed by multiplying by the 'rate' to give a 'price'. Once prices for all the components have been calculated, the total for the task is determined by addition.

## DRIVEWAY COSTING EXAMPLE

Consider another example, one that raises questions about just how the work will be done. In this example, the cost of installing a sub-base is calculated.

The project involves constructing a rectangular driveway, 13m × 3.6m. This task consists of excavating the area to a depth of 200mm and then putting in a layer of crushed stone, to be 100mm thick, as a sub-base. This is a DIY project, so no allowance will be made for labour costs.

### Excavation

$$13m \times 3.6m = 46.8m^2$$

To expect the excavation and the sub-base to be identical in size to the finished driveway is naïve – it is standard practice to allow for some 'spread' of the sub-base and 'working room' for the excavation, so an allowance of 250mm will be made. One of the 3.6m edges represents a garage and therefore the adjusted calculation now reads:

$$(13.0m + 0.25m) \times (3.6m + 0.25m + 0.25m)$$
$$= 13.25m \times 4.1m$$
$$= 54.3m^2$$

The excavation needs to be 200mm deep, so:

$$54.3m^2 \times 0.2m = 10.86m^3, \text{ say } 10.9m^3$$

Excavated earth has a density of around 1.9 tonnes per cubic metre (t/m³), so, the 10.9m³ dug out on this project weighs in at a not-insignificant 20 tonnes or more. How will this be handled? Will a mini-digger be used or is it going to be spades and wheelbarrows? And how will it be removed from the site? A mini-digger seems to be the best option, but there is a minimum three-day hire, at £70 per day for the digger, plus £60 for delivery and collection. That is £270, but the machine will be onsite for three days; it should only take a day to complete the excavation work, so the mini-digger could help to spread and level the crushed stone.

As for disposal, skips are the only viable option. A call to the local skip hire company reveals that each skip will cost £125, with no time limit on the length of use, that each will be capable of holding around 4.5m³ of excavated material, and that a permit from

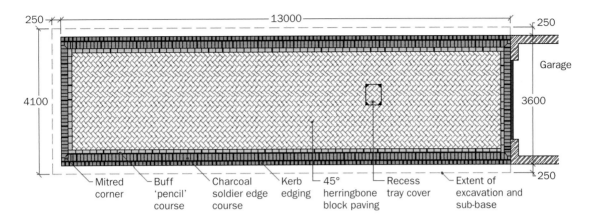

250 — 13000 — 250

4100

Garage

3600

250

Mitred corner — Buff 'pencil' course — Charcoal soldier edge course — Kerb edging — 45° herringbone block paving — Recess tray cover — Extent of excavation and sub-base

*Layout of the example driveway being costed.*

the local authority will be needed if the skips are to be 'parked' on a public highway. The permit costs £25 and can be issued for any length of time, but the council require that all skips are marked with temporary road signs, traffic cones, and, if left overnight, they must be illuminated with orange roadwork beacons. Luckily, the skip company will supply all of these items for an additional £45.

So, for the disposal:

$$10.9m^3 \text{ excavated} \div 4.5m^3 \text{ skip capacity}$$
$$= 2.4 \text{ skips}$$

– but try as you may, getting hold of 0.4 of a skip is virtually impossible, so a total of three skips will be required. The 0.6 of a skip spare capacity will surely be filled with other debris during the course of the works (assuming that the neighbours do not first fill it with their rubbish as soon as your back is turned).

## Sub-base

The sub-base is to be constructed from a crushed stone known as Type 1. The density of this material is approximately $2.2t/m^3$, and the cost is either £25 per tonne when it is delivered in individual bulk bags, or £14 per tonne as a direct delivery from the quarry in a tipper wagon. However, there is a 'part

load charge' of £8.50 for each tonne not carried on a 20t wagon.

The extended excavated area is $54.3m^3$ and the sub-base will be 100mm thick:

$$54.3m^2 \times 0.1m = 5.43m^3 \text{ @ } 2.2t/m^3 = 12t$$

12t in bulk bags costs:

$$12 \times £25 = £300,$$

or 12t direct delivery costs:

$$12 \times £14 = £168,$$
plus the part-load charge:
$$(20t – 12t) = 8t \times £8.50 = £68,$$
giving a total of £168 + £68 = £236.

It has been decided to use a geotextile membrane between sub-grade (the excavated ground) and sub-base. It comes in $30m^2$ packs, each costing £25; two will be used, and so this item will also need to be added to the costing.

That is all the materials accounted for, and, as mentioned, the labour is zero-costed because it is a DIY project. However, a vibrating plate compactor will be needed to consolidate the sub-base once it has been spread and levelled.

## Totting-up

The overall costing can now be completed:

| Item | Quantity | Rate (£) | Total (£) |
|---|---|---|---|
| labour | 1 | 0 | 0 |
| *plant* | | | |
| mini-digger: | | | |
|    hire | 3 days | 70/day | 210 |
|    delivery/collection | 1 | 60 | 60 |
| plate compactor: | | | |
|    hire | 1 week | 35/week | 35 |
|    delivery/collection | 1 | 20 | 20 |
| *skips* | 3 | 125 each | 375 |
| licence | 1 | 25 | 25 |
| safety kit | 1 | 45 | 45 |
| *materials* | | | |
| DTP1 | 12t | 14/t | 168 |
| part-load charge | 8t | 8.50/t | 68 |
| geotextile | 2 rolls | 25 each | 50 |
| **Total** | | | 1,056 |

The principles outlined in these two examples illustrate the procedure used to cost any paving project. The work is broken down into the three components of labour, plant and machinery, the necessary quantities are calculated, rates are applied and the total arrived at.

Examples of typical costing sheets are given on pages 73 and 74. The rates are averages and should not be taken as indicative for any particular area or region. Check local suppliers for up-to-date rates. The first costing sheet prices up all the work required for the driveway used as an example in this section. The second looks at the much simpler costing of a straightforward rectangular patio.

*OPPOSITE: Costing sheet for the driveway.*

| | | Quantity | Rate | | Sum | | Item Totals |
|---|---|---|---|---|---|---|---|
| 1.1 Excavation [1] | 200 mm depth | 10.9 m³ | | | | | |
| Labour | 2 men for 1 day | 2 days | £ 120.00 /day | £ | 240.00 | | |
| Cart Away | 4.5 m³ skips | 3 nr | £ 125.00 ea | £ | 375.00 | | |
| Excavator | | 2 days | £ 70.00 /day | £ | 140.00 | | |
| Delivery/Collection | | 1 nr | £ 60.00 ea | £ | 60.00 | | |
| | | | | | | £ | 815.00 |
| 1.2 Geo-textile | 50m² roll | 1 nr | £ 35.00 ea | £ | 35.00 | | |
| Labour | | 1 hr | £ 15.00 /hr | £ | 15.00 | | |
| | | | | | | £ | 50.00 |
| 1.3 Sub-base [1] | 100 mm thick | 12.0 T | £ 14.00 /T | £ | 168.00 | | |
| Part-load charges | | 8.0 T | £ 8.50 /T | £ | 68.00 | | |
| Excavator | | 1 days | £ 70.00 /day | £ | 70.00 | | |
| Plate Compactor | | 1 wk | £ 35.00 /wk | £ | 35.00 | | |
| Labour | | 4 hr | £ 15.00 /hr | £ | 60.00 | | |
| | | | | | | £ | 401.00 |
| 1.4 Kerbs | 125x125x100 in red | 26 m | £ 8.30 /m | £ | 215.80 | | |
| Mortar | | 0.73 m³ | £ 110.00 /m³ | £ | 80.44 | | |
| ST1 Concrete bed and haunch | | 1.1 m³ | £ 95.00 /m³ | £ | 108.68 | | |
| Labour | | 6 hr | £ 15.00 /hr | £ | 90.00 | | |
| | | | | | | £ | 494.92 |
| 1.5 Soldier Courses | 200 mm wide | | | | | | |
| Charcoal Blocks [2] | 6.64 m² exactly | 7.0 m² | £ 9.20 /m² | £ | 64.40 | | |
| ST1 Concrete bed | | 0.50 m³ | £ 95.00 /m³ | £ | 47.31 | | |
| Labour | | 4 hr | £ 15.00 /hr | £ | 60.00 | | |
| | | | | | | £ | 171.71 |
| 1.6 Soldier Courses | 100 mm wide | | | | | | |
| Buff Blocks [2] | 3.16 m² exactly | 3.5 m² | £ 9.20 /m² | £ | 32.20 | | |
| ST1 Concrete bed | | 0.50 m³ | £ 95.00 /m³ | £ | 47.31 | | |
| Labour | | 2 hr | £ 15.00 /hr | £ | 30.00 | | |
| | | | | | | £ | 109.51 |
| 1.7 Bedding Sand | 50 mm thick | 4.8 T | £ 22.50 /T | £ | 108.86 | | |
| Blockwork [2] | 37.20 m² exactly | 39.1 m² | £ 9.20 /m² | £ | 359.35 | | |
| Labour | | 37.2 m² | £ 8.00 /m² | £ | 297.60 | | |
| Cutting-in | | 30.8 m | £ 2.50 /m | £ | 77.00 | | |
| Block Splitter | | 1 nr | £ 35.00 ea | £ | 35.00 | | |
| Jointing | 40 Kg bags | 4 nr | £ 3.50 ea | £ | 14.00 | | |
| | | | | | | £ | 891.82 |
| 1.8 Recess Tray | 600x450 | 1 nr | £ 35.00 ea | £ | 35.00 | | |
| Mortar | | 1 nr | £ 10.00 ea | £ | 10.00 | | |
| Labour | | 2 hr | £ 15.00 /hr | £ | 30.00 | | |
| | | | | | | £ | 75.00 |
| | | | **Sub-total** | £ | 3,008.95 | | |
| Overheads/Profit | 20 % | | Add | £ | 601.79 | | |
| | | | **Nett Total** | £ | 3,610.74 | | |
| VAT | 17.5 % | | VAT | £ | 631.88 | | |
| | | | **Gross Total** | £ | **4,242.62** | | |

| | | | |
|---|---|---|---|
| Total cost of materials | £ | 1,769.35 | plus VAT |
| Total cost of plant | £ | 340.00 | plus VAT |
| Total Labour charges | £ | 899.60 | plus VAT |
| | £ | 3,008.95 | |
| **Rate per m²** | £ | **90.65** | |

[1] includes 250mm spread to entire perimeter
[2] allows for 5% wastage

Patio is to be 5.5m x 4.3m in Indian sandstone with tumbled clay paver edging

| | | | Quantity | Rate | | Sum | | Item Totals | |
|---|---|---|---|---|---|---|---|---|---|
| 1.1 | Excavation [1] | 100 mm depth | 2.9 m³ | | | £ | 120.00 | | |
| | Labour | 2 men for ½ day | 1 days | £ 120.00 | /day | £ | 120.00 | | |
| | Cart Away | 4.5 m³ skips | 1 nr | £ 125.00 | ea | £ | 125.00 | | |
| | | | | | | | | £ | 245.00 |
| 1.2 | Soldier Courses | 200 mm wide | | | | | | | |
| | Tumbled clay pavers [ | 3.92 m² exactly | 4.2 m² | £ 21.25 | /m² | £ | 89.25 | | |
| | ST1 Concrete bed | | 0.29 m³ | £ 95.00 | /m³ | £ | 27.93 | | |
| | Labour | | 3 hr | £ 15.00 | /hr | £ | 45.00 | | |
| | | | | | | | | £ | 162.18 |
| 1.3 | Bedding Sand | 50 mm thick | 2.8 T | £ 22.50 | /T | £ | 63.86 | | |
| | Cement @ 8:1 | 25 Kg bags | 15 bags | £ 3.20 | /bag | £ | 48.61 | | |
| | Raj flagstone [2] | 19.89 m² exactly | 20.9 m² | £ 17.50 | /m² | £ | 365.48 | | |
| | Labour | | 19.9 m² | £ 15.00 | /m² | £ | 298.35 | | |
| | Pointing Mortar | 1 bag mix | 3 nr | £ 10.00 | | £ | 30.00 | | |
| | Pointing Labour | | 3 hr | £ 15.00 | ea | £ | 45.00 | | |
| | | | | | | | | £ | 851.29 |
| 1.8 | Step Units (risers) | 200x450 | 8 nr | £ 24.25 | ea | £ | 194.00 | | |
| | Mortar | 1 bag mix | 2 nr | £ 10.00 | ea | £ | 20.00 | | |
| | Labour | | 3 hr | £ 15.00 | /hr | £ | 45.00 | | |
| | | | | | | | | £ | 259.00 |
| | | | | **Sub-total** | | £ | 1,517.47 | | |
| | Overheads/Profit | 20 % | | Add | | £ | 303.49 | | |
| | | | | **Nett Total** | | £ | 1,820.97 | | |
| | VAT | 17.5 % | | VAT | | £ | 318.67 | | |
| | | | | **Gross Total** | | **£** | **2,139.64** | | |

| | | | |
|---|---|---|---|
| Total cost of materials | £ 964.12 | plus VAT | |
| Total cost of plant | | plus VAT | |
| Total Labour charges | £ 553.35 | plus VAT | |
| | £ 1,517.47 | | |
| **Rate per m²** | £ 45.72 | | |

[1] includes 250mm spread to entire perimeter
[2] allows for 5% wastage

*Costing sheet for the hypothetical patio.*

# CHAPTER 6

# Tools

There are a few basic tools that are required on almost any hard-landscaping job. These might be found in many half-decent DIYer's sheds, but there are also a few pieces of kit that are peculiar to our trade and may have to be hunted down. Let us start with the essential digging kit.

## DIGGING TOOLS

A spade is the single most important item in the digging kit. As well as for excavation, it is used to shift bulk materials such as crushed stone and sand, to level them out, and to mix mortar and concrete. A garden or border spade will suffice for smaller jobs, but the spade most commonly used by contractors is a 'taper-mouth'. This features a blade that becomes wider as it moves up from the cutting edge, making for easier digging, but still with adequate breadth to move the excavated material. Note also how the blade is angled, relative to the shaft, so that shovelling from a flat surface is made easier.

While the spade is ideal for digging in soft ground, it needs some help to cope with stony ground or hard clay, and so a pick (pick-axe) or mattock is highly recommended. Most picks have a point at one end of the head with a chisel at the other. The broader the chisel, the more effective it will be in clayey ground, while the point can be reserved for the really hard stuff.

The final essential item is the wheelbarrow. There is plenty of choice, but it is hard to find a better barrow than the standard builder's barrow: a tray pressed from steel and fixed to a tubular steel chassis with a pneumatic tyre. Flimsy, folded and galvanized bodies, thin steel tube for the chassis, and a solid wheel are to be avoided at all costs. The average garden barrow may be good enough for moving a few bedding plants or clearing leaves in autumn, but it is unlikely to last more for than a few hours when being battered with heavy earth, crushed stone or paving blocks.

## PAVER'S KIT

The paver's kit comprises those tools used primarily to assist in laying the paving. A rake is used to level out granular materials, such as the sub-base stone and the bedding sand; avoid lightweight garden rakes and look instead for a 'tarmac rake' featuring a wider head, sometimes with stays attached to the shaft. The whole is fixed to a chunky wooden stele

*Digging kit: sledge hammer, tapermouth spade, pick, shovel, rake and barrow.*

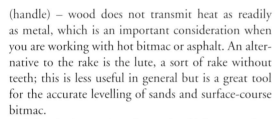

*Laying kit: a landscape rake, a rubber maul and the obligatory stiff brush.*

*Block splitters come in many shapes, sizes and colours but usually rely on a lever to close the upper and the lower blade.*

(handle) – wood does not transmit heat as readily as metal, which is an important consideration when you are working with hot bitmac or asphalt. An alternative to the rake is the lute, a sort of rake without teeth; this is less useful in general but is a great tool for the accurate levelling of sands and surface-course bitmac.

When laying paving of a regular thickness, such as small element flags and block pavers, the bed is usually prepared by 'screeding', which involves scraping off an area of sand or other granular material to create a smooth and accurate bed. This is achieved by using a screeder board, which is essentially a 3–4.5m length of dead straight timber or aluminium, along with a set of screed rails, which are normally 20–25mm-diameter steel tubes more commonly used as electrical cable conduit.

To consolidate the paving, a maul (mall) is the tool of choice for larger pavings such as flags. Traditionally, these were large wooden items, with the head held together by iron bands. These are still used in some parts of the country, but the compound rubber maul has largely replaced them. These are less prone to splitting and rather more forgiving of the heavy-handed flagger who has not learned to temper his own strength. The best rubber mauls are reinforced with fibres, usually cotton string, which helps to bind the rubber and limit the bounciness, but these are becoming increasingly hard to find.

Small rubber mallets are used with smaller pavings. These typically have a wooden shaft that carries the rubber head, which weighs around 1kg. There are smaller versions, but these do not have sufficient wallop to consolidate paving bricks or small flags effectively. Beware of mallets in which the shaft does not fully pass through the head but relies on the head being fixed to it by means of a small tack. These have a tendency to split where the tack pierces the rubber.

A block splitter is also an essential piece of paving equipment, although it is possible to manage without one if a power cut-off saw is available. However, splitters are fast and simple to use and do not produce a cloud of dust in the process. On the downside, they can be awkward to use in the wrong hands, and are suitable for use only with concrete block pavers and small element flags. Special splitters are available for cutting clay pavers, and there are hydraulic models for cutting larger flags, natural stone flags, cropping setts and the like, but these are not really everyday kit.

A good, general-purpose block splitter will be able to cope with units up to 450mm in length or width, and 100mm in thickness. The blades will be adjustable and replaceable, and the cropping lever will be extendable to give more power when cutting.

## HAND TOOLS

Essential hand tools include a hammer, a selection of chisels, trowels and floats, string lines, line pins, a tape measure and a spirit level.

## Hammers

The best all-round hammer is the lump or club hammer, 1–2kg of cold steel atop a wooden or fibreglass shaft. Choose a hammer that feels balanced in the hand. The lighter, 1kg (2lb) hammers are better for general use; the heavier 2kg (4lb) versions are more suitable for experienced mason paviors using them to crack stone flags and road kerbs. A brick hammer is useful as its chisel end can be used to trim the cut edges of blocks, bricks and flags, but generally, at around 0.5kg, it is too light for all-round sitework.

## Chisels

Chisels come in all sorts of shapes and sizes, but the two most commonly encountered forms are the cold chisel and the bolster. A cold chisel is essentially a short, hexagonal or octagonal bar of hardened steel with one end ground to form the characteristic chisel end. It is ideal for breaking into hard materials, such as brick or concrete. The bolster is a specialized chisel that has its cutting edge broadened out to 50mm or more, and is often used to cut pavings, although its best use is for removing snots (small, surplus lumps of mortar or concrete) from floors and slabs.

Two other types of chisel used by mason paviors are the point and the pitcher (pitching chisel). These were essential tools in the days before power saws, but they are a rarity nowadays. The point, as its name implies, comes to a point and is used to start holes in stone. The pitcher is used to dress a face on to walling stone, or to cut flags, both concrete and stone.

## Trowels and Floats

Trowels and floats constitute another group of tools that come in a wide array of formats and perform an equally wide range of tasks in the hard-landscaping trade, from fine-tuning the bed for a flagstone, to laying bricks. If only one trowel was to be chosen, then a brick trowel would be the best option. These are 200–350mm in length, but the key is to find a trowel that feels comfortable in the hand and that the user is capable of lifting when it is fully laden with mortar. The shorter, narrower brick trowels are best suited to DIY-ers, while contractors and bricklayers tend to go for the longer and wider versions.

A pointing trowel is a much smaller version of the brick trowel and is used for fine detail work, such as pointing and benching (levelling and smoothing small concrete fillets). A pointing bar is useful when it comes to tooling (smoothing off) mortar-pointed joints, although one can be formed in the shed from a short length of copper pipe.

Larger areas of concrete, and even bedding layers of sand, are levelled and smoothed using a float, of which the most useful for the paving trade is the

*Hammers and chisels: rubber mallet, cold chisel, scutching hammer, pitching chisel, lump hammer, bolster chisel.*

*Trowels: brick trowel, pointing trowel, pointing bar flat, pointing bar V, pointing bar half-round, and large float.*

larger, bull-float, but there are also specialist floats, such as the arris trowel that is used to form a curved or bull-nosed edge on concrete slabs.

## String Lines and Line Pins

String lines are used to create guides to alignment and level. There are mason's lines, bricklayer's lines, and plain, ordinary builder's lines, which will be adequate for most residential paving jobs. The string itself is twisted cotton, nylon or, increasingly, polypropylene, and is of a finer grade than that used to tie up parcels and the like. All string stretches when extended over a run of a few metres, but the better quality lines stretch less and are also less prone to snapping and breaking than some of the cheaper ones.

Line pins are used to support the string lines. Site pins are usually of steel, formed from 10 or 12mm bars and 600 to 1200mm in length, although, for DIY and residential works, bamboo canes or dowelling may be used. The advantages of steel pins are that they are more robust, less likely to snap and have an almost indefinite working life. Many contractors form their own line pins by sawing lengths of steel reinforcement bars into more convenient lengths.

*Measuring: 30m open frame tape, pocket tape and a spirit level.*

## Measures

A typical contractor will have half-a-dozen or more tape measures, but there are two that should be regarded as essential: the pocket tape, which is usually 3m or 5m in length and is marked in millimetres and/or eighths of an inch, and the open-frame tape, which is usually 10–100m in length with centimetres or half-inches as the smallest sub-division. The pocket tape is used for accurate measuring work, such as marking the size of a flag to be cut, while the larger tape is used for setting out. Most steel pocket tapes are satisfactory, but some of the budget products have weak or sub-standard recoil springs and often go rusty as soon as you breathe on them. Ensure that there is a locking mechanism on the tape that prevents it from recoiling until required. With the larger tapes, fibreglass is a good choice since it is rustproof, light and flexible. An open frame is a better choice than a case as they are less prone to twisting, and, when the tape itself is wet or dirty, it is easier to rewind.

## Spirit Levels

Finally, we come to the spirit levels. Again, this is a tool that comes in many forms and an average tradesman will have at least two: a small 'boat' level for checking the levelling of small units, and a longer 'beam level', which is used to check the levelling of string lines and larger runs of paving. The accuracy of a spirit level is dependent on its length: a longer level is more accurate than a short one, assuming that the vials are accurately set in the first place. Most contractors rely on a 1200 or 1800mm beam level as much as possible, using the 200–300mm boat level when the beam level is just too cumbersome. A simple check of accuracy is to use the level to set up a surface that reads as flat, with the bubble positioned as centrally as possible within the vial. Then, rotate the level through 180 degrees and the bubble should read exactly the same. If it is closer to one end of the vial, then either the surface has shifted or the level is inaccurate. You must decide which is more the likely.

## POWERED TOOLS

There is no doubting that powered tools have made construction tasks much easier than they were just a generation ago, but how many are actually needed for a typical residential paving project?

## Excavators

Having a mini-excavator and a hydraulic skip loader may be a boon, but they might not be necessary, or it might not be possible to get them in to where they are needed. While many contractors now have their own mini-excavators, hiring-in as and when needed is still a popular option. A 1 to 1.5t excavator is usually adequate for simple, shallow-dig projects, but for bigger areas, or those projects requiring deeper excavations, a larger machine, up to 3t or so, would be a better choice. On big jobs, a tracked or wheeled, full-size excavator is essential.

Do not worry that an excavator is not being used every minute of the working day. Even if it is used for only a couple of hours, it is still a sound investment since it will do the work of ten men in that short period.

## Power Saws

The power saw has rendered many traditional cutting techniques obsolete, although some of the more naturalistic materials, such as stone flags or setts, may not look 'authentic' when trimmed to perfectly straight lines with the aid of a modern saw. The two most popular versions are those with a self-contained petrol engine and those powered by electricity. The first can be used anywhere, whereas the second need to be plugged into the mains or to a separate generator.

When there are only a few cuts to be done, then an angle grinder can usually cope with the work, but the small diameter of the blades and the proximity of the operative's hands to the cutting edge mean that this option should really be considered only for the smallest projects and the occasional repair.

There are two main types of blade used with power cut-off saws (and angle grinders). The abrasive blades may be only a fraction of their cost, but they have a reduced cutting time when compared with even mid-range diamond blades. Contractors use diamond blades, and usually the mid-range or top quality blades rather than budget offerings. Diamond blades have exceptional durability, making each linear metre of cut cheaper, pro rata, than cuts made by using an

*Power saw fitted with diamond-tipped blade. The notches help to cool the blade when it is in use.*

abrasive blade; but, more importantly, they are faster at cutting and they give a consistent depth of cut, unlike the cheaper abrasive blades which wear down to nothing with use.

## Plate Compactors

The vibrating plate compactor is often referred to by the name of one of the better manufacturers, Wacker. As all vacuum cleaners are commonly referred to as 'Hoovers', so plate compactors are generally known as 'Wacker plates', although the name is often misspelt as 'Whacker', in the mistaken belief that the name comes from their use to 'whack' loose materials until they are compacted. The real reason for the name is that Wacker Gmbh in Germany pioneered the plates.

As should be expected, there are many of different brands and types, but for residential paving work, a paver plate or a plate generating a force of around 12–15kN will be fine. Additional extras include a urethane 'mat' or 'sole' that can be fitted to the underside of the plate to minimize damage to the pavers when they are consolidated, and removed again when compacting sub-base material or something similar. The models offering reverse gears and/or remote control are a little extravagant for most residential jobs but are increasingly used on larger sites.

Although plate compactors are used to consolidate bitmac, they are really not the best tool for anything but small areas and minor repairs. For the best finish with bitmac, a smooth-drum vibratory roller is the tool of choice.

# CHAPTER 7

# Pavement Layers

## INTRODUCTION

The aim of this chapter is to introduce the basic concepts in pavement construction, providing definitions for technical terms, along with the underlying principles, so that the reader will be in a position to understand the project they are undertaking or be armed with sufficient understanding of the jargon to avoid being misled by any less scrupulous contractors.

All paving consists of a number of layers and the fundamental principle is that each successive layer builds on those beneath to create a structure that improves in terms of firmness and resilience, accuracy of levels and (finally) appearance, as the layers build up. Some types of paving have only two or three layers, while others consist of half-a-dozen or more. In general, more individual layers make for a stronger pavement, but for drives and patios this is not always the case: a single layer of concrete laid 100mm thick is usually much stronger than a layer of patio flags on a bed of sand and cement. Each layer has its own characteristics, and an understanding of these layers, along with their differences and similarities, will help to ensure better construction practices.

## SUB-GRADE

This is the term used to describe the bare earth, the ground, but only once it has been prepared, which means stripping it of all vegetation and topsoil. In the trade, this preparatory excavation is

*The sub-grade is 'reduced' by digging down to formation level.*

normally described as 'reduce to formation level'. This means removing the surface vegetation, whether it is grass and weeds or trees and shrubbery, and then excavating and removing topsoil, which has a relatively high organic material content, to reveal a more stable sub-soil or clay beneath, and digging down to a level from where the actual construction will start. It is because organic material degrades and decomposes over time, resulting in the settlement of anything lying on top of it, that all such material has to be removed before pavement construction can begin.

Again, using sweeping generalizations, stripping the vegetation and the associated root zone removes around 50mm or so from the ground. The depth of topsoil varies enormously from place to place, but is usually 75 to 250mm thick. The sub-soil beneath often has a more brownish tan colouring, a higher clay content, and is generally firmer and harder to dig. This is what is needed for the base of a typical patio or light-use driveway. The ground surface above which all subsequent layers are constructed is referred to as the 'formation level'.

Ideally, when a site is reduced to formation level, the excavated surface would be relatively even, with no significant high spots or low points, and would generally follow the profile of the proposed pavement, so that each of the build-up layers can be fairly regular in thickness. However, any soft spots or suspect ground should be excavated and removed – having a sound, reliable formation level is more important than having a flat or even, regular surface. It is better to remove too much than to leave behind an area that can, and probably will, compromise the long-term performance of the completed paving.

## SUB-BASE

This is a vital layer and often the first major structural one of a typical pavement. It forms a flexible but strong intermediate between the underlying sub-grade (or any improvement or capping layer) and the base or laying course materials above. Sub-bases are not always used with patios, but they are essential for most types of driveway paving, particularly block-paved driveways. The single most common cause of settlement or channelization of a block-paved driveway is an unsuitable, unsatisfactory, or non-existent sub-base.

When explaining driveway structure to householders, I would often compare the sub-base layer to that of the underlay of a carpet. Some contractors would claim that a sub-base would not be necessary because the existing ground has been there for years and would not be going anywhere. However, the same might be said of a floor in the house, and it does not matter how much is spent on the carpet – the surface layer that is actually seen and trafficked – if the underlay, the sub-base if you will, is not present and properly laid, using the correct quality of material, then even the finest Axminster will never feel right.

Sub-base materials come in three main forms: unbound, cement-bound, and bitumen-bound. For most residential paving requiring a sub-base, an unbound material is used, although for projects where there may be exceptional loads, or where the sub-grade is unreliable for whatever reason, a cement-bound material (referred to as CBM) may be used. Bitumen-bound sub-bases are virtually unheard of for residential driveway and patio paving.

Unbound materials for sub-base construction are commonly some type of crushed rock or, increasingly in this age of recycling, a crushed concrete. The basic principle is that the blend of larger particles and fines (the sand- or dust-sized material) can be combined by careful mixing, distribution and compaction to create a firm, void-free layer of reliable material. Further, the stratified nature of even a relatively shallow sub-base results in any point loads being spread over a larger area, thereby increasing the ability of the pavement to support heavy weights. This is achieved by passing on the downward force of any load to a successively greater number of particles. Imagine a point load exerting pressure on a single, large particle in the sub-base: this particle may rest on two other large particles, so the pressure is shared or halved between the two. These two particles, in their turn, rest on others, which rest on others, and so forth, so that the load is rapidly spread over a larger area. Obviously, this is a much-simplified explanation of the physics involved, but the essence remains correct: sub-bases help to spread or dissipate point or narrow loads over a wider area.

## DTp1

Most builders' merchants, and a growing number of the larger DIY suppliers, offer a sub-base material of some description. The 'best' is referred to as Type 1, DTp1, MoT1, 804, 40 mil down, and probably a few other names specific to particular regions. For patios and many pathways exactly which material is used is not critical, but for driveways, especially those carrying heavier-than-average vehicles such as 4×4s, then it is best to play safe and go for the genuine article.

## Other Materials

Other materials supplied for and used as sub-bases include the following.

### Ballast

This is a mix of gravels and fines, often from an old riverbed quarry or a marine dredging operation, which is particularly popular in the relatively aggregate-poor south-east of England. There tends not to be too much control over the precise quantities of fines or gravels, it is a case of you get what you are given, even if it is particularly sandy. This would be a problem on civil or commercial projects, but for patios and driveways, it is of less consequence, and when the only alternative is to import a crushed rock from possibly 100 miles away and at twice the cost, ballast is often the only realistic option.

### Crusher Run

This is a term that refers to any rock that has been run through a crushing process. The crushed material is usually sieved to separate it into size ranges, and so there is 50mm crusher run, where the largest particle size is, as the name suggests, 50mm. All sizes below 50mm will also be present, hence its alternative name of '50mm to dust', but there is no strict control over the proportion of the smaller, finer particles included in any particular batch; 75mm crusher run, 100mm and 150mm are also produced. For many driveways, and most patios that require a sub-base, 50mm crusher run is a suitable material, provided that it has a reasonable blend of lumps and fines, so that there are not many voids or 'holes' in the surface when it has been compacted.

### Quarry Waste/Scalpings

These are a pair of 'hit-and-miss' terms that cover all the shards, off-cuts, dust and sweepings-up that are generated as part of any normal quarrying operation. The type of rock being quarried, and the type of product being produced have a direct bearing on the quality of the waste and/or scalpings. Quarries producing flagstones often generate a relatively 'clean' or dust-free waste, while those producing a crushed aggregate often sell off the rubbish, the low-quality detritus, and the dusty tailings, as a general purpose 'fill' material that may be eminently suitable as a capping or improvement layer but may not be up to scratch for a structural sub-base. There is so much variation in what is sold as quarry waste or scalpings that it is simply not possible to give a definitive statement regarding suitability: some quarry wastes are superb materials for driveway and patio projects, while others are fit for little more than backfilling; only by examining and assessing the quality of the rock type and the ratio of fines to larger particles is it possible to judge whether it is worth parting with your money for this material.

### Planings

This is the name given to the chewed-up-and-spat-out road surfacing that is generated as part of the more-or-less endless maintenance of our main roads and motorways. A huge beast of a machine tracks along the old, worn carriageway, scratching at the surface with great tungsten teeth, rasping it, milling it, and reducing its level so that it may be overlaid with a new surface course, or replaced all together by new base (binder) and surface-wearing courses. The waste produced is sent up a conveyor belt and into the back of a wagon that carries it away. It is sometimes used as a low-quality improvement layer for private trackways or farm access roads, but it is also sold as a so-called sub-base material for private driveways. Planings present a number of problems: although the material tends not to contain too many large lumps and seems to have a reasonable blend of fines and larger particles, there is no control over the exact blend, and some batches are very dusty while others are almost clean, with only a little or no fines content. The type of surface that was planed also has a direct affect on the usefulness of the planings –

some road surfaces are stone-rich macadam and therefore the planings will have a high content of natural, inert rock, while others are mostly asphalt, which is much softer and has a tendency to degrade or break down as it ages.

As with the quarry waste and scalpings described above, there is so much variation with planings that it is not possible to give a definitive indication of their suitability, but, as a general rule, they are best avoided for driveway use. Some planings, notably those from asphalt surfaces, have been observed to act like a fluid when saturated with water, and this can result in definite ruts or channelization of the pavement after a few years. Leave planings for the farmers looking for a cheap, gravel-like material to top up their pothole-ridden, suspension-shattering access lanes. Planings may be cheap and readily available, but in many cases there is too big a question mark over their long-term performance to warrant the risk.

*Hardcore*

To round off this section looking at alternative sub-base materials, we come to the near universal term for these materials: hardcore. The word is nearly always in the top twenty search terms used to find information on the Pavingexpert website, and it amuses me greatly to think of those disappointed surfers expecting something quite different from crushed rock, broken concrete and half-bricks! Hardcore is another of those loose terms applied to construction materials. Some people will use the term when they are referring to a premium aggregate such as DTp1, and some to refer to old bricks, broken flagstones, bits of stone, crumbled mortar and goodness knows what. And that is the problem: one man's hardcore is another man's pile of useless, fit-for-nothing rubble.

Generally speaking, the old bricks and broken flagstones type of hardcore is not suitable as a sub-base material. The 'particles' are too large, there are too many voids left in the material when it is compacted, and some of the components are barely capable of supporting their own weight, let alone that of a pavement and a car. However, if it has been crushed and blended, as is happening more and more as the trade looks to recycle and reuse more waste products, then it might be ideal. The same require-

*Sub-base layers work by spreading the load over a wider area.*

ments as previously mentioned should be sought: an inert, crushed material with no pieces larger than 40 to 50mm, exhibiting a good blend of larger particles and fines that will give a 'tight' surface when compacted.

## GEOTEXTILES

This is another item on the typical patio or driveway shopping list that has almost as many names as it has uses. Geotextiles, geosheets, geomembranes, separation membrane, geofabric, and even the incorrect 'plastic sheet' are all terms that have been used.

For residential paving, the main function of these products is as a separation membrane; that is, as a layer of fabric that prevents the mixing of those materials above and those below. There is some debate as to whether the separation should be between sub-grade and sub-base, or between sub-base and laying course. There are arguments to be made for both, but for residential paving, my vote goes for the separation of sub-grade and sub-base. By keeping the sub-base intact, by preventing any squishy sub-grade material from being 'pumped' into the sub-base, by preventing the sub-base from being lost into the sub-grade, the integrity of the whole structure is maintained. To me, this seems more of a pressing need than the need to prevent bedding course material trickling down into any small voids within the sub-base.

But that is not to say that a geotextile is needed for every job. They have their uses, but the majority of home paving projects do not warrant the inclusion of a membrane. If the sub-grade is sound, dry and firm, then the risk of mixing between sub-base and sub-grade is minimal, given the relatively light loads imposed on the paving. However, if the sub-grade is problematic, if it is made-up earth or sticky clay, or moving in any way when uncovered, then a geotextile can dramatically improve the performance of the pavement as a whole. It is hard to conceive just what effect a simple sheet of non-woven fabric can have, but in the twenty-five years or so the construction industry has been using these products their worth has been proved time and again. Even a mid-strength geotextile sheet laid across ground of uncertain quality and covered with 100–150mm of good sub-base material can be sufficient to take the weight of a wagon, and if they are good enough for our road builders, they are more than good enough for driveways.

A couple of caveats, however: these construction membranes are not the same as the landscape fabrics that are sold in garden centres and DIY stores. Those products are often a woven fabric and are intended to separate loose mulches, such as bark, gravel or slate chippings, from the earth beneath, or to ensure that no weeds can grow in protected areas within a greenhouse or some other horticultural building. Although they are, technically speaking, a separation membrane, they are not designed to be a structural component of a pavement. Secondly, these geo-textiles are often wrongly sold by paving contractors as 'weed-proofers'. The claim is made that, by including one of these membranes within a pavement, weeds will not be able to grow. While not quite being an outright lie, this is stretching truth to the limit. It is true that weeds cannot grow through these membranes, but then weeds cannot grow through a properly constructed pavement. If a driveway consists of 150mm of well-compacted sub-base material, topped with 40mm or so of well-compacted grit sand and then 50–60mm of solid paving, very few weeds would be capable of growing through that. There are some super-weeds, such as Mare's Tails (*Equisetum spp*) and Japanese Knotweed (*Polygonum spp*) that can manage it, but your average dandelion, daisy or rosebay willowherb will never manage it. Consequently, unless you have one of these pernicious weeds, to use a geotextile for weed protection within a pavement is usually a case of over-engineering. Further, most weeds afflicting a pavement do not grow *through* the paving, they grow *into* it. The annoying weed seeds settle into a joint or a crack, or in detritus that has accumulated on the surface, and send down adventitious roots to establish themselves. Even in what was clean sand there are enough nutrients to be gleaned and scavenged to

*A geotextile separation membrane helps to keep the sub-base intact and so reduces the risk of settlement and improves the load-bearing capacity of the pavement.*

render a typical paving joint an attractive home to a wandering dandelion seed. How does a geotextile sheet buried 100–300mm beneath the surface of the paving prevent these seeds from germinating and the weeds from establishing themselves?

To summarize: geotextiles are a boon if there is any uncertainty regarding the suitability of the sub-grade, but the correct type must be used. And although the use of a membrane can dramatically strengthen a pavement, it will be totally ineffective against surface weeds.

## BASE COURSES

Base courses are stiffer, tougher, less flexible construction layers that are more commonly used with heavy-duty pavements or tarmacadam surfaces, where they may also be known as binder courses. A base course is rarely used beneath a patio, unless there are specific and exceptional reasons to do so, although one may be used beneath a driveway constructed from stone flags, setts, cubes or cobbles.

There are two principal types of base course: flexible and rigid. A flexible base course is usually constructed by using bitumen-bound materials (bitmac) which may not seem flexible to the average person, but is so defined by civil and structural engineers since it remains capable of minor flexion and movement. Perhaps semi-rigid (or semi-flexible) would be a better descriptive term. A rigid base course is usually constructed from concrete, and is therefore inflexible, solid and incapable of accommodating minor movement.

Base courses are typically used when constructing rigid pavements. The base acts as an intermediate layer, between the unbound and totally 'flexible' sub-grade and/or sub-base beneath, and the rigid bedding and surfacing layers above. The degree of rigidity required often dictates whether a flexible, bitmac base or a fully rigid, cement-bound base would be the more suitable. Where the surface layer is to be flexible (for example, a tarmac surface or wearing course), then the base would be flexible. Where the surface course is to be rigid, such as a mortar-bedded and jointed sett pavement, the base may be flexible or rigid. Where heavy loads are expected, a rigid base is the more likely.

A general rule of thumb is for flexible surfaces to be constructed on flexible bases and rigid surfaces to be constructed on rigid bases, paraphrased to 'flexi on flexi: rigid on rigid'.

## KERBS AND EDGINGS

These are the linear paving elements used at the perimeter of a pavement or at the interface between two paved areas with different intended uses. As mentioned previously, there is no hard and fast rule about what constitutes a kerb and what an edging, but to generalize, a kerb usually has 'upstand' or 'check', that is, part of the kerb rises proud above the regular paved area and offers 'check' to the passage of a wheeled vehicle. An edging is usually 'flush', laid at the same level as the paved surface and offering no check to wheeled traffic. Both kerbs and edgings constrain a paved area, that is, they frame it and hold it in place.

With some kerbs and edgings this constraint is an essential part of the pavement structure, while with other structures it may be purely decorative, and both functions may occur within the same pavement. Consider a typical block-paved driveway: the edge courses laid against the garden or lawn will need to function as constraining or restraining edges, while those edge courses against the brickwork of the property are included purely for aesthetic reasons, their function is to provide 'balance' to the finished look of the driveway, to complete the 'framing effect' of the edge course as a whole.

Most kerbs are modular units, laid on a bed of concrete with additional concrete placed at the rear of the laid units to hold them in position. This is known as 'haunching'. The only notable exceptions are the extruded or slip-formed kerbs that are constructed from asphalt or concrete to the required shape, style and length as part of the construction process. These are not particularly common for driveways in Britain and Ireland, although they are used in Australia and the United States.

Obviously, the most popular use of 'kerbs' is to prevent traffic (either pedestrian or vehicular) straying from a pavement, running over the edge or travelling on to a paved area deemed unsuitable or unsafe. A kerb is often used to keep cars off a garden area or a designated footway, and the type of traffic expected often determines the most suitable type of kerb. Cars and bicycles on low-speed driveways can be kept in their place by using a simple kerb with only a few centimetres of check, while on busy roads or places where HGVs may be present, 'high containment kerbs' providing 400mm or more of check and haunched with steel-reinforced concrete may be necessary.

Edgings may also be modular units, laid on and haunched with concrete. However, there are also a number of plastic, aluminium and steel products that are used, as well as the traditional timber edgings, which rely on pegs, nails or spikes driven into the ground to keep them in place, along with the weight of the earth behind them. These are often adequate for light-use driveways and patios where the ground is firm enough to support the pegs or spikes, but they are not normally recommended for use on busy driveways or where the underlying ground is unstable in any way.

## BEDDING

The bedding or laying course material is the 'cushion' or accommodation layer between modular units used as the surface or paving layer and those layers lying beneath, such as the sub-base or base layer. Its purpose is to provide support for the paving units and to enable variations in thickness or size of the paving materials to be accommodated without upsetting the finished paving level.

A wide range of materials is used as bedding, but, again, there are two main types: bound and unbound. The type of paving being laid usually determines the most suitable form of bedding. Referring back to the rule of thumb given for base courses, rigid surfaces usually rely on bound bedding materials, while flexible paving relies on unbound bedding.

Unbound bedding is usually some form of granular aggregate. Sand is the most popular, particularly a coarse or grit sand, although crushed rock, grit or stone dust may also be used. The key requirement is that the material should be fine enough to be accurately levelled out without leaving significant voids, yet not so fine that it retains or

soaks up water. A good and simple test for the suitability of a bedding material is to douse it with clean water, allow it to drain for two minutes and then grab a handful – when it is squeezed, no water should trickle out between the fingers. This indicates that the material is sufficiently free-draining to be used as a bedding material.

Most builders' merchants and the larger DIY stores stock a coarse grit sand for use as a bedding material for paving. It may be known as grit sand, coarse sand, sharp sand, concreting sand, bedding sand, Class M or Zone 2, among other, more regional names. Building sand, which is also known as soft sand, mortar sand, fine sand, plastering sand or masonry sand, is *not* suitable as a bedding material because it tends to retain too much water and often has a relatively high silt and clay content. The average grain size is larger with a grit sand than is the case with a soft sand. A grit sand contains grains of up to 3–4mm in diameter, while most building sands are composed of grains of 2mm or smaller. This grittiness should be obvious when you look at the sands, and certainly when rubbing them between your fingers.

When it comes to bound bedding, the two usual materials are involved once again – bitumen and cement. Bitumen-bound bedding material is somewhat rare in Britain and Ireland, although it is used in the USA. We seem to prefer cement-bound materials, and these fall into two categories: mortars and concretes.

## MORTARS AND CONCRETES

What is the difference between mortars and concretes? It is necessary to generalize, but a mortar essentially comprises fine aggregates (sand) mixed with cement and water, while a concrete consists of coarse aggregate (gravel), along with the fine aggregate, the cement and the water. There is a sort of continuum from a fine mortar based on soft sand, to a coarse mortar made from grit sand, a fine concrete that incorporates 6mm aggregate, a medium concrete that contains 10–20mm gravel, and all the way through to a coarse concrete with 20–40mm gravel.

The job in hand usually determines which is the most suitable cement-bound material to use, and the thickness of the bedding layer is critical in this decision. Obviously, a 20mm medium concrete would be no use where a 10mm bed is required, and, similarly, a fine mortar based on a 2mm sand would be wasteful if you are trying to create a 50mm deep bed for setts or cobbles.

All mortars and concretes vary in two principal factors: the cement content and a characteristic known as 'slump', which simply means the wetness or 'sloppiness' of the material. The cement content largely determines the end strength of the mortar or the concrete – the more cement is included, the stronger the mortar/concrete. Most mortars and concretes used for paving and hard-landscaping will contain 10 to 25 per cent cement.

Slump also affects the final strength of a concrete or a mortar since there is an optimal ratio for the cement and the water content. Mortars and concretes that are wetter or drier than the optimum will be weaker, although in most cases a materials testing laboratory would be needed to prove this. If you are really interested, the optimal water:cement ratio (known in the trade as the w/c ratio) is around 0.5, which means that there should be roughly 1kg of water (which is, by definition, the same as 1ltr) to every 2kg of cement.

For mortars, a typical slump is known as a wet mix. The mortar is capable of standing in peaks, and is wet enough to flow (known as 'plastic') when squeezed, as when laying bricks, but not so wet that it cannot support its own weight and therefore collapses and levels out. The key test is to 'slash' the mortar, that is, to cut it with the point of a trowel to a depth of 30–50mm. The mortar should remain 'slashed'; the slash should not close up due to the flow of the mortar under the effect of gravity, but should remain open.

## How Wet?

There are five commonly used slumps for mortars and concretes in the paving trade and each has its specific uses, advantages and disadvantages. These are summarized below.

**Slurry mix** A mortar or fine concrete prepared to a soup-like consistency. This type of mix is most commonly used to wet-grout flags, setts, cubes and

| *Useful Mortar and Concrete Mixes for Paving Work* | |
|---|---|
| **Mix** | **Ingredients** |
| Class II general purpose mortar | 3–4 parts soft sand<br>1 part cement |
| Bedding mortar for flags | 8–10 parts grit sand<br>1 part cement |
| ST1 concrete | 6 parts 10mm gravel<br>3 parts grit sand<br>1 part cement |
| ST4 concrete | 4 parts 10mm gravel<br>2 parts grit sand<br>1 part cement |

cobbles, where the slurry mortar is brushed over the surface and allowed to flow into the open joints, filling any voids between and beneath the paving units. The surplus slurry is then removed from the surface, usually by sweeping.

**Wet mix** As described above, a wet mix is one prepared with an optimal water content. Wet mixes are possibly the most widely used as they are ideal for tasks such as bricklaying and pointing, and are a popular choice for bedding, although the fluid nature of the mix may sometimes be more of a hindrance as it allows previously laid units to 'float' and shift in level when adjacent units are tapped down on to the wet bed.

**Moist mix** This is a much stiffer version of the wet mix, with just sufficient water added to ensure that the aggregates and cements are bound together. The usual test for a moist mix is to squeeze a handful tightly and ensure that no surplus water trickles out between the fingers, as described earlier. As the cement in such a mix is 'active', this must only be done when wearing suitable gloves. Moist mixes are sometimes used for pointing because they are not quite as messy as wet mixes, but their primary use is as a bedding material as there is sufficient moisture present to allow the material to bind and bond to the paving units, but not so much that the fluidity of the bedding becomes a problem.

**Semi-dry mix** This is a very stiff mix that relies on the natural moisture within the sand and/or the aggregates to initiate the hydration (hardening) of the cement. There is no water added to the mix unless the aggregates are exceptionally dry. Some moisture needs to be present to start the hydration of the cement, but it is kept to a minimum. Semi-dry mixes are primarily used for bedding. The low water content means that they remain workable for relatively long periods (4–6hr is not uncommon), there is no problem with fluidity when units are laid, and there tends to be no adhesion of the paving to the bedding, which is why semi-dry mixes are preferred for tasks such as laying kerbs and some flagstones.

**Dry mix** This is normally used only for jointing. A dry, or almost dry, sand is mixed with cement and then brushed into open joints where, it is hoped, the dry powder will draw in moisture from the bedding or a pre-existing wet mortar 'butter' and so start to set. The advantage is that, as a dry(ish) powder, there is little risk of staining the surface, and so this is often a popular choice for grouting patio flags where a clean finish is often essential. However, as there is little control over just how much moisture will be drawn into the dry mix, it may sometimes cure in a haphazard manner and end-up as crumbly, rather stale yet friable, and nowhere near strong enough to

---

**Be Careful!**

Cement is nasty stuff. When mixed with moisture it is strongly alkaline and can 'burn' the skin, eating away at the tissue, resulting in a painful condition that persists for several days. Gloves should always be worn when working with cement or materials containing it, such as mortars and concretes. Bare arms and legs should be covered up since any direct contact between skin and cement risks dermatitis, irritation or burning. Safety goggles are recommended if there is any risk at all of cement powder or splashes getting into the eyes.

---

last more than a winter or two as a jointing material. Sometimes, contractors 'water-in' a dry jointing mix by sprinkling water on to it from a watering can. While this does ensure that there is some moisture present to improve the hydration of the cement, it can have the effect of 'splashing' the dry, cement-containing powder on to the now-damp surface of the paving, and nullify the original benefit of no staining.

## PAVING OR SURFACE LAYER

This is what it is all about – the uppermost surface, the one that everyone can see, the one that carries the traffic, incorporates the design and determines the finished look; it is the cream on the trifle of layers beneath, the icing on the cake.

As we have already seen, there are a great many possible paving or surfacing materials that can be used for this final layer, and, because there is such a wide range, it is not easy to see what they have in common. However, all paving materials have three key qualities: they should be reasonably attractive, they should be relatively smooth, but not glassy or slippery, and they should be fairly hard-wearing. No one wants to construct a pavement that is unattractive (even though some of the driveways and patios we have all seen may tempt us to believe otherwise). Ugly or unattractive paving is, more often than not, a result of the wrong materials in the wrong place and laid in the wrong way. Just what constitutes an attractive paving unit or surfacing material is clearly a matter of personal taste, but the general trend is for relatively fine-grained materials, with a flattish surface that allows for the comfortable passage of feet and wheels, and perhaps a touch of colour. This can consist of flagstones, patterned concrete or tarmac – all quite different materials yet they meet the description above. Flagstones have a tight and relatively smooth surface, as does patterned concrete, while the tarmac used for the surface course is usually a fine-grained material, with 6mm being the largest size particle typically used for driveways (10 or 14mm is sometimes used for public highways).

That paving materials are not slippery stems from the need for safety. Our feet and our wheels need some traction, some grip, if they are to travel across a

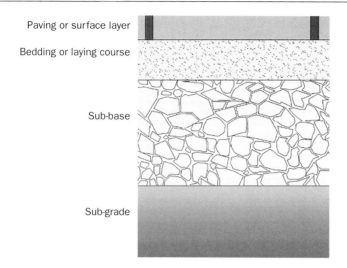

*The main construction layers used in residential paving work.*

surface and so we seek a balance between what looks good and what offers a safe footing. Smooth, polished materials are not overly popular as a surface or paving layer, although they do exist. Paving materials are tested for slipperiness: aggregates are tested for a characteristic known as PSV – polished stone value – which is a measure of how slippery an aggregate or stone becomes when repeatedly trafficked and 'polished', while manufactured units such as concrete flags and block pavers are assessed for PPV – polished paver value, a similar concept to PSV, but applied to the surface of a whole unit rather than to a component aggregate. Materials not intended for use in public areas, such as patio flags, may not be assessed for PSV or PPV, but generally, most manufacturers and suppliers are keen to ensure than any paving products they provide will not constitute a hazard nor prove to be treacherous or dangerous under normal usage.

Durability is important because no one wants to use a material that will wear away or abrade in next to no time. All paving materials do abrade, it is a perfectly natural process, but the better materials tend to be much more hardwearing than their cheaper counterparts. Granite, basalt and Yorkstone are very hardwearing, while some of the wet-cast concretes used for patio flags are relatively soft and so tend not to be used in high-traffic areas. Concrete block pavers are somewhere in between: they are not as hard nor as abrasion resistant as is, say, a granite cube or sett, but they are tougher than a typical patio flag. This makes them excellent for driveways, where a typical wet-cast patio flag would have its surface worn away and the concrete interior exposed in a relatively short time.

These different materials are laid in a variety of ways, and this is what will be considered in a later chapter.

# CHAPTER 8

# Setting Out

This is the first task when starting the physical work on any paving project. The area to be paved needs to be established and marked out on the ground partly to ensure it 'fits' within the space available and partly to ensure that it is balanced, even, and in proportion with everything else. It is much easier to adjust the length and alignment of a chalk line than it is to move an already laid kerb edge. Setting out also helps when it comes to visualizing what falls and gradients will be involved, if these have not be pre-determined during the design and planning stage. Setting out generally relies on a number of fairly simple geometric principles. If you ever sat through a boring maths class wondering what possible use Pythagoras would ever have in your life, you are about to find out. Setting out uses straight lines and triangles, circles and arcs, and combines them to ensure that

*Setting-out kit: marker pins, chalk line, string line, spray paint, tape measure, and spirit level.*

the finished pavement will actually be a rectangle rather than a trapezium (a lop-sided rectangle, you will no doubt recall).

## BASIC TOOL KIT

A few tools will be needed:

- tape measures, preferably two, with one being at least 10m in length;
- spirit level, preferably 1200mm in length;
- string line, mason's or bricklayer's line is preferred to parcel string;
- marker pegs or steel line pins;
- lump or club hammer to drive in pins and pegs;
- chalk, crayon and/or spray paint to aid marking;
- calculator (not always needed).

If an automatic or laser level is available, it can be a great boon, but many simple residential driveways and patios can be set out without them.

## SETTING OUT THE SHAPE

The usual procedure when setting out is to first establish the lines, the curves and the overall shape of the pavement, and leave establishment of levels, the height or the depth of the finished surface at any given point, until later. This helps to ensure that, when the levels are established, they are exactly where they are required and not in 'the general vicinity of' or 'close by'. This helps to ensure greater accuracy and a better finish for the completed pavement.

The best way to explain how setting out is done is

to work through examples. An idealized pavement has been drawn up and is illustrated on page 92. This driveway incorporates many of the features that may be encountered on a typical residential pavement, and will illustrate how straight lines, perpendiculars, parallels, arcs, S-curves and diagonals are established on the ground. This exercise emphasizes how important it can be to have a good plan before starting. It need not be a work of art, lovingly detailed and intensively annotated – a simple sketch, even one that is not to scale, is just as useful, provided that it contains all the necessary information.

## THE EXAMPLE DRIVEWAY

The proposed driveway, as shown on page 92, will run almost the full length of the front of the house, except for 3m at the southern end, which is lawn and will have a series of stepping stones leading to the garden shed. There is a pair of quarter-circle planter beds on either side of the entrance porch. At the north end of the house there is a small return path leading to the rear gate and the single garage block.

The access to the drive from the public highway will be via a pair of gate pillars that have yet to be built. The threshold itself is not parallel to the house, but the driveway will be set out square to the property rather than to the public highway. The northern section will link to the garage by an S-curve, while on the south side there is a 45-degrees cut-back linking to a curve of 3m radius. The southern gate pillar at the entrance to the drive, and therefore to the southern edge of the driveway, is to align with the north-facing side of the porch.

### Establishing a Perpendicular

The first setting out task will be to establish a perpendicular line (defined as a line that is at a right angle or 90 degrees to a base line) from the front of the house, labelled A next to the porch, out to the public highway, labelled as point B, in the diagram on page 94.

*Using a 3–4–5 Triangle*
The simplest way to establish a perpendicular is to create a right angle, and the simplest way to do this over any distance of more than about 1m is to create

a right-angled triangle. One of the most serendipitous truths of the universe is that, if a triangle has sides that are in the proportions 3:4:5, then it must contain a right angle. The triangle may have sides that are 3cm, 4cm and 5cm, or 30m, 40m and 50m – it will always contain a right angle. Even if the sides are multiples of 3:4:5, say 15ft, 20ft and 25ft, it will contain a right angle. This simple fact has been used by architects, masons and builders for thousands of years and is enshrined in that favourite from everyone's schooldays, Pythagoras' theorem. This states that the square of the length of the hypotenuse in a right-angled triangle is equal to the sum of the squares of the lengths of the other two sides. The hypotenuse in a right-angled triangle is the longest side, opposite to the right angle, so if we square its length (multiply it by itself) and then do the same with the other two sides and add them together, we should end up with the same value. So …

$$5^2 = 4^2 + 3^2$$
$$5 \times 5 = (4 \times 4) + (3 \times 3)$$
$$25 = 16 + 9$$
$$25 = 25$$

We can use this to construct a 3–4–5 triangle from the front of the house. From the porch at point A, it is approximately 10.5m to the end of the house. It is best if the '4' side of the triangle lies against the house, as this means that the shortest side, the '3', is the one that will be projected out from the building, and by using the shortest side, the possibility of error is minimized.

We could use a length of 10m for the 4 side along the house; 10 is 2½ multiples of 4, so this would mean that the 3 side would need to be 3 × 2½ or 7.5m and the 5 side (the hypotenuse), will be 5 × 2½ or 12.5m. A tape measure is fixed at point A and a distance of 10m is measured along the wall of the house and marked on the wall as point C. It is worth noting that all measurements are done at, or very close to, ground level. If one point were to be taken at ground level and another at, say, 1m high up the wall, there is a degree of vertical displacement that would result in a slight, but possibly serious, error.

While the end of the measuring tape is established at point A, the 3 side of the triangle, which is to be

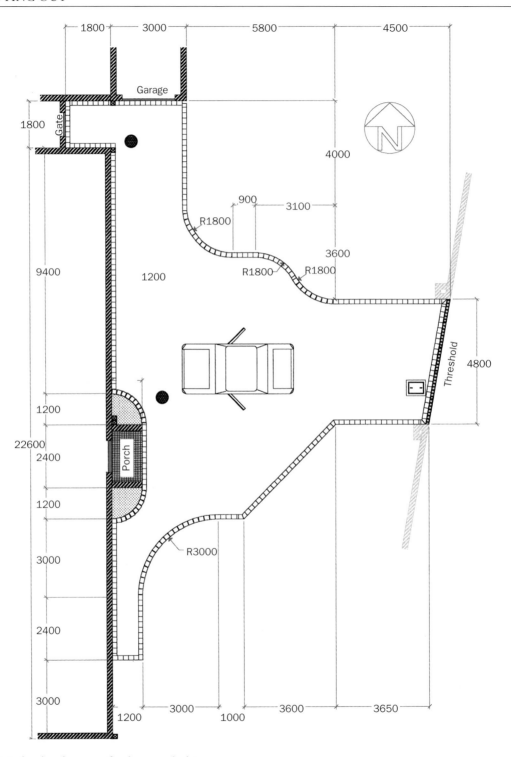

*This is the plan drawn up for the example driveway.*

7.5m in length, can be set out. So, with the tape firmly anchored at A, an arc is marked on the ground at point D. An arc has to be used since we do not know exactly where point D lies, other than that it must lie on an arc of 7.5m centred on A. The arc could be scratched on to the existing surface, marked by a crayon or chalk, or painted with an aerosol can of temporary marking spray.

Moving the tape to point C (or using a second tape measure), we can now establish an arc of radius 12.5m to define the hypotenuse. Where the two arcs cross, the point of intersection, is the third point of our 3–4–5 triangle. We will drive in a marker pin at this point, ensuring that the pin is vertical; 12mm steel road pins, or 1m lengths of steel reinforcing bar are ideal for this.

A line from point A to point D is the perpendicular required to determine the position of the left-hand gate pillar. However, the line we have created is only 7.5m in length and the gate pillar is approximately 5m further on. This line can be extrapolated, that is, extended to the required point, and this is best done using a string line. One end of the string is fastened at A, the line is paid out past the pin driven in at D and all the way up to the public footpath. By pulling this string line taut from a position on the footpath, we create a straight line, and then by adjusting the alignment of the string line, moving it from right to left, until it just touches the marker pin at D, we can establish point B on the edge of the footpath, on the perpendicular line from A, through D. This defines the position of the proposed gate pillar.

From this line, we can now set out the southern side of the driveway. Consulting the original plan, we can see that this southern edge is to extend for a distance of 3650mm from the gate pillar. If we tension and then fix the string line at point B, we can measure back towards A and at 3650mm, mark point E with another steel pin.

## Establishing a 45-degree Line

The next point to establish is F, the other end of the 45-degree line from E. This can be done be creating a square, with the upper right-hand vertex (corner) at E and the lower left-hand vertex at F. The diagonal of the square will be the line E–F. (See diagram on page 95.)

Measuring back 3600mm towards the house along the A–B line is simple enough to do, so we can do this and mark point G. From G, swing an arc by using the tape measure and mark the arc where we think point F will be. Next, we need more Pythagoras: we know that two sides of the imaginary triangle E–F–G are each 3600mm, so

$$(EF)^2 = 3600^2 + 3600^2;$$

we shall need the calculator for this one:

$$(EF)^2 = 12,960,000 + 12,960,000 = 25,920,000;$$

by using the calculator's square root facility, we find that

$$\sqrt{(25,920,000)} = 5,091.2;$$

so the line EF should be near enough 5091mm; we can swing an arc from point E, with a radius of 5091mm (or as near as we can manage) and mark the arc so that it intersects with the arc swung earlier from point G; where the two arcs intersect is our point F and a marker pin can be driven in here.

We can check the position of point F. The drawing shows that F should be a distance of

$$1200mm + 3000mm + 1000mm = 5200mm$$

from the front wall of the house. If we fix a tape at point F and swing an arc of 5200mm towards the wall of the house, it should just touch the wall, if our setting out (and the drawing) is accurate.

Using extra checks such as this can help to make sure that the setting out work is as precise as is possible. It may emerge that the arc swung from F towards the wall is 50mm short of making contact. Should we move point F or accept that there is nearly always some inaccuracy when setting out a project such as this? For most residential paving jobs, an accuracy of ±50mm is pretty good, and, in this example, it is probably best to leave F where it is. If it were found that F was out of position by 100mm or more, then further checks should be made, perhaps by calculating further triangles and using Pythagoras to check on the position of several points.

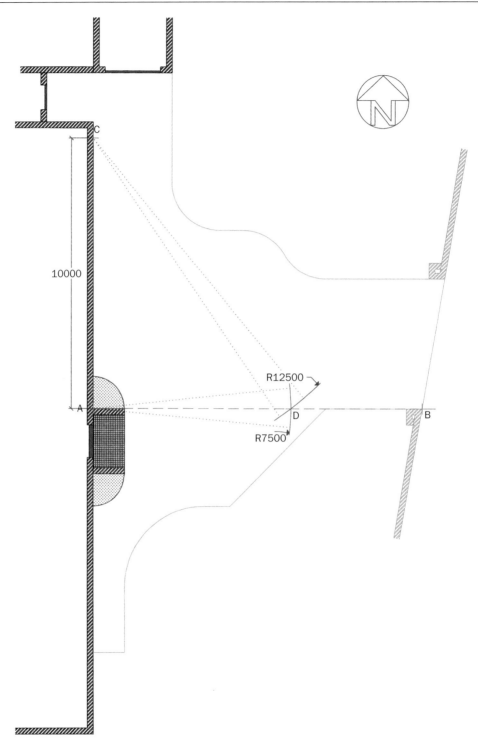

*Setting out the first perpendicular line using a 3–4–5 triangle.*

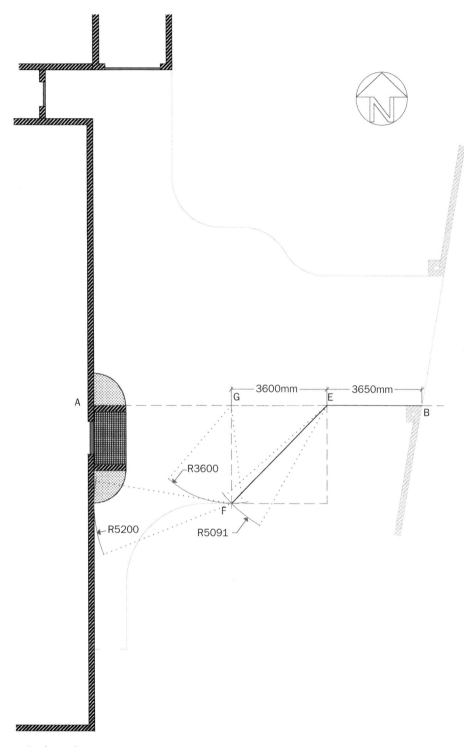

*Establishing a 45-degree line.*

## Further Setting Out

Assuming that F is actually in the right position, we can continue the setting out of the south side. (See diagram on page 97.) The next point to establish would seem to be point H, which is just 1m back towards the house from point F. Point H could be established by creating a right-angle triangle F–G–H and by using Pythagoras once again to calculate the hypotenuse G–H. Alternatively, we could create another perpendicular from the house wall out to F and then measure off the 1m back to H. What is probably the best method, given the potential inaccuracy of point F, would be to create a series of perpendiculars from the house wall, which have been labelled J–H, K–L and M–N, and use these to determine the remaining points on this side of the driveway.

Point O, the corner of the building, is a firm fixed point and an easy one from which to measure. The distance from O to M is, according to the drawing, a rather convenient 3000mm. This could be used as a base line for another 3–4–5 triangle. Measuring from point O will give point M, which can be marked, and we shall use this as one side of the triangle. It would be silly not to use it as the 3 side; so that means that we need to swing an arc of 5000mm from point O and one of 4000mm from point M, and the two will intersect at point P. A string line can then be stretched from M out to P, a distance of 1200mm measured from M along this line, and that gives point N, which can be marked with a steel pin.

Point L is

$$3000mm + 2400mm = 5400mm$$

from the corner of the house, and can be marked when measured. This length is a multiple of 3 (1800 × 3), so, if an arc of

$$1800mm \times 4 = 7200mm$$

is swung from point L, and an arc of

$$1800mm \times 5 = 9,000mm$$

from point O, we have yet another 3–4–5 triangle. The arcs are swung, they intersect at Q, a string line

is stretched from L to Q, a distance of 1200mm along this string line is measured from L, and we now have point K, which can also be fixed with a pin.

Finally, point J is 3000mm from point L, so we can repeat the 3–4–5 triangle, swinging an arc of 4000mm from J and one of 5000mm from L. They should intersect at R, and when a string line is stretched from J to R, it should be possible to extrapolate it through to F, established earlier. We can then measure 1000mm from F or 4200mm from J, or measure both, and this will give point H, which is also marked with a steel pin.

The final setting out task on this side of the driveway is to establish the centre of the 3000mm arc linking point H to K. This has been labelled as point I and is referred to as the 'origin' of the arc. Points H and K are known as 'tangent points'. This means that they are the points at which the arc ends and the straight lines begin. In geometric reality, the arc continues to form a circle, and a tangent point is the point where the circle just touches the straight line.

We could set out the origin, I, in a number of ways. We could simply measure

$$1200mm + 3000mm = 4,200mm$$

from the wall along the perpendicular line L–Q. We could swing 3000mm arcs from points H and K, and they should intersect at point I. If we do both, we can be sure that we have got the actual origin of the required arc nailed down fairly accurately.

We now have all the necessary reference points on the south side of the driveway. We have seen how perpendiculars may be created by using a 3–4–5

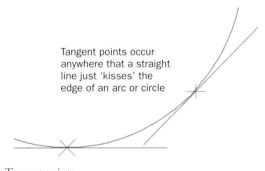

Tangent points occur anywhere that a straight line just 'kisses' the edge of an arc or circle

*Tangent points.*

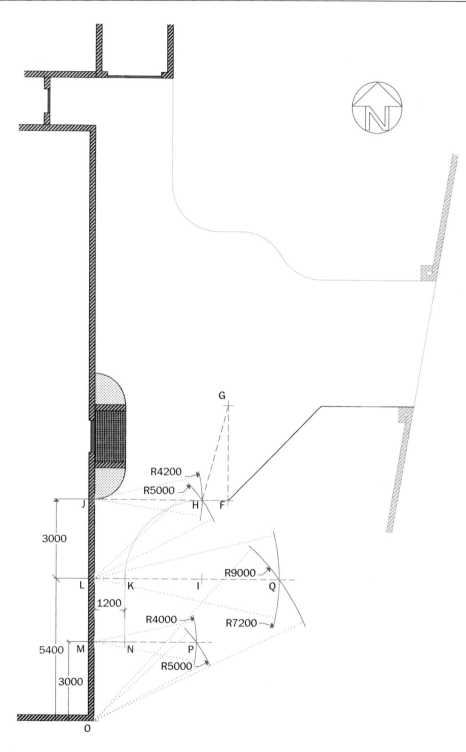

*Further setting out.*

triangle, how lines can be extrapolated to determine points lying outside such a triangle, how Pythagoras' theorem can be used to create other right-angle triangles, and how tangent points, origins and arcs are related. We shall build on this learning and add to it as we set out the north side.

## Setting Out the North Side

This side consists of fewer points. Working back from the end nearest the garage, there is a tangent point S, 4000mm from the corner of the garage, and then an arc of 1800mm radius linked to a short section of 'straight', U–V, which then links to an S-curve, V–X–Y; that brings us to the final length of straight, Y–AA which takes us to the position where the northern gate post is to be.

For those sections closest to the house it is easiest to set out from the front wall; but for the position of the gate post it is probably best to establish line Y–AA as a parallel to line B–E on the south side and allow any discrepancy to be lost within the S-curve where it is unlikely to be noticed.

Therefore point S will be established first. (See diagram on page 99.) The garage is 3m wide (CC–DD), so this could be used as one side of a 3–4–5 triangle. An arc of 4m radius can be swung from CC, and a 5m arc from DD, which should intersect at S. We can double-check the position of S as it should be exactly 3 metres from the front wall of the house at EE, which should be 4 metres from the other corner of the garage at DD. It is essential that the edge CC–S is parallel to the front of the house, and, because the quadrilateral CC–DD–EE–S is a true rectangle (two pairs of sides of equal length and four angles of 90 degrees), the two diagonals CC–EE and DD–S should be identical when measured. This comparison of crossed diagonals is the key check for parallel in a setting out exercise.

Once checked, a pin can be driven in at S and then a string line stretched from EE past S to point T, which should be 1800mm beyond point S and will be the origin of the first arc, S–U. A further pin can be established at point T, and this can then be used to set out the arc S–U.

If we go back to the front wall of the house, we can measure back from point EE a distance of 1800mm to establish point FF, and then by swinging an arc of

$$3,000\text{mm} + 1,800\text{mm} = 4,800\text{mm}$$

from this point, this should intersect with the 1800mm arc centred on point T at the tangent point U.

We could extrapolate the line FF–U through to and beyond where point Z should be, but it is more important for finished appearances that the arc centred on Z aligns with the right-hand gatepost and the edge of the driveway, Y–AA. So, for now, we shall extrapolate the line FF–U, take it beyond where we expect point Z to lie, and leave it at that for the present. Once we have established the right-hand gatepost and that edge of the driveway, we should be able to check back to the line FF–U and see just how accurate our setting out, or the original site plan, has been.

So, moving on to establishing the line Y–AA (see diagram on page 101): this could be done in a number of ways. A 3–4–5 triangle A–GG–HH with sides of 4800mm (A–GG: 3 × 1600mm), 6400mm (GG–HH: 4 × 1600mm) and 8000mm (A–HH: 5 × 1600mm) will give point HH that can be extrapolated. Alternatively, or in addition, we could use the line B–E established earlier as the base for a 3–4–5 triangle, which would establish point BB. This would be achieved by measuring 3000mm along B–E and marking the point JJ, from which a 5m arc can be swung which should intersect with a 4m one swung from B at point BB. The same dimensions may then be used to create a second 3–4–5 triangle E–KK–Y. A string line stretched from Y–BB should align exactly with the line AA–GG that may have been established previously. One final check would be to use Pythagoras on the rectangle B–E–Y–BB, with sides of 3650mm and 4800mm, then the two diagonals, B–Y and E–BB should measure:

$$\sqrt{(3650^2 + 4800^2)}$$
$$= \sqrt{(13,322,500 + 23,040,000)}$$
$$= \sqrt{(36,362,500)}$$
$$= 6030.1 \text{ or } 6030\text{mm rounded off.}$$

Checking that these two diagonals measure the same

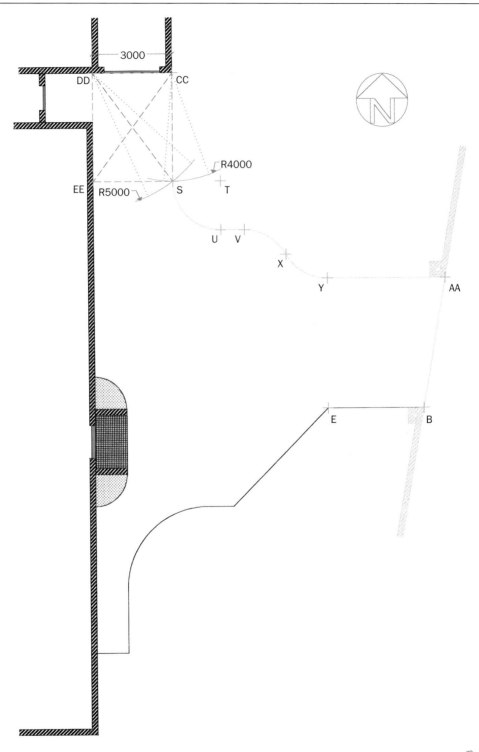

*Setting out the garage perpendicular.*

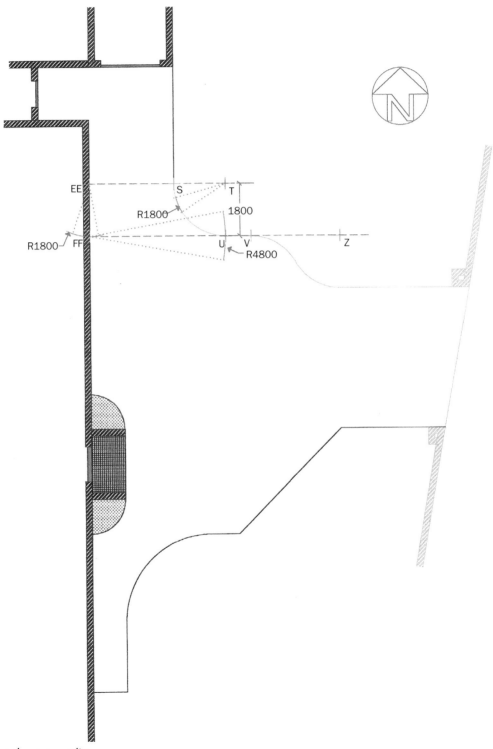

*Setting out the garage radius.*

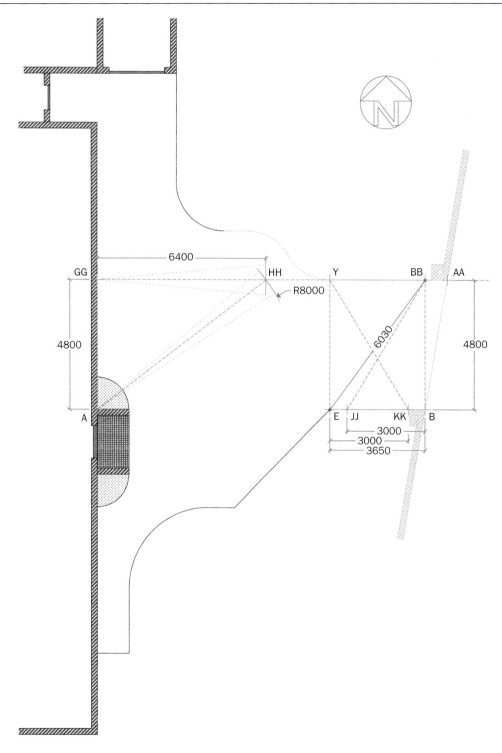

*Setting out north-side edges.*

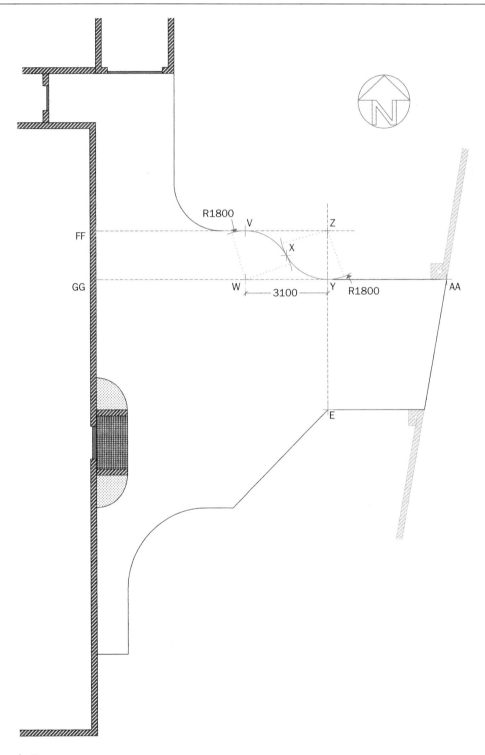

*Setting out the S-curve.*

is the final proof needed that the line B–E must be parallel to line BB–Y.

The final part of the setting out is to establish the arc centred on point Z (see diagram on page 102), which can be found by extrapolating the line E–Y and measuring 1800mm beyond point Y. If the line FF–V has been left in place, point Z should lie on the intersection of that line with E–Y.

The arc can now be marked by fixing a tape to point Z and swinging an arc of 1800mm, marking the ground in the process. The final arc, centred on point W, can be created by measuring 3100mm along line GG–AA from point Y. The arc so created should just touch the arc centred on Z at point X, and then come around to point V, which lies on the line FF–Z, approximately 900mm from point U established earlier.

## Finishing Off

Congratulations if you have followed so far without developing a headache: you have managed to complete the setting-out process for our idealized driveway. The key strategies to remember from this exercise are the use of 3–4–5 triangles to create right angles and perpendiculars, the use of crossing diagonals within a rectangle to check for parallel, and the importance of basic geometry, particularly triangles, rectangles and arcs, to the setting out process.

The other key fact to bear in mind at all times is that most setting out is a matter of compromise. There is an adage that if you were to get ten surveyors to measure a line on a site, you would get eleven different results. No matter how accurate we try to be, there will always be some error. Perhaps the tape was sagging slightly, or twisted, or the wind was catching it when the measurements were taken, or maybe the spray paint marks used were slightly wider than normal and this introduced an error. Any discrepancy of up to 50mm can normally be excused and explained away by site variation, but if it were noticed that the discrepancy were greater than 50mm, this would suggest that something was amiss. Either the site survey and drawing were incorrect in some way, or there was a measuring error or miscalculation somewhere. The only way to be sure is to check and double check the previous setting out

work and the drawing dimensions to determine just where the error(s) arose.

*Compromise*
Whatever the cause of minor discrepancies, compromise will be necessary, so it is important to know where these compromises can be 'lost'. On this project, it is essential that the edges of the driveway from B–E and AA–Y are parallel since any significant discrepancy (30mm or more) would be evident when viewed from the house or the public highway, especially if a square block paving or other regular pattern of paving were to be used. Similarly with the edges S–CC near the garage and K–N on the south side of the porch. These, too, are rectangles, and any deviation from parallel would probably be noticeable. Any compromise or minor discrepancy could more easily be 'lost' or 'made up' on the arcs, particularly the *S*-curve on the north side, or on the 45-degrees cut-back on the south side.

*Working Room*
At this stage, we should have a series of marker pins established at specific, critical points on the layout (see diagram on page 104). It is important to realize that most of these points represent the edge of the planned driveway. When it comes to excavation and preparation, we need to ensure that there is adequate working space, allowance for spread, room for haunching and so on. Those pins positioned at edges should be driven in to a depth of at least 450mm, so that, when the excavation work is done, as long as not more than 250mm is removed from the existing surface, the remaining 200mm of pin embedded in the ground should be about adequate to support the pin.

On most driveway and patio projects, allowing 200–300mm of 'spread' at the edges of the excavation should give sufficient space to accommodate the spread of a sub-base and haunching for edge courses and/or kerbs, so it would be sensible to mark the ground wider than the pins as a guide to excavation. This is best done with sand lines: fill an empty bottle with a dry sand and then gradually pour it out to create a line on the surface. A sand of a colour that contrasts sharply with the existing surface works best, and any mistakes or errors are easily scuffed out and redone.

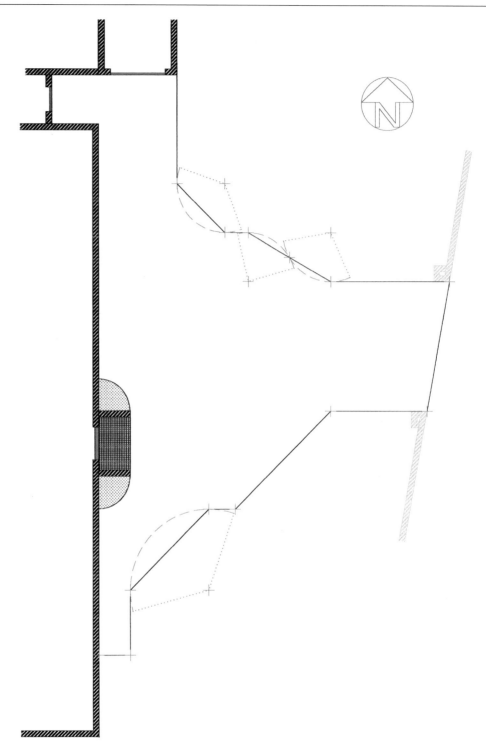

*All key points set out and marked with pins.*

One final point to make before moving on is that the few pins so far established are unlikely to be sufficient to accurately guide the eventual paving work to a correct alignment. This is most obvious on the arcs and the *S*-curve, where all we have at the moment are the origins and tangent points. Additional marker pins would be a boon when it comes to laying the paving, but, at this stage, they would simply be cluttering the site and so are best left out until the excavation is complete and the sub-base is in place.

## SETTING THE LEVELS

So far, all the setting out work we have done has been about establishing line, defining where edges will lie, where arcs will swing, and where corners and other vertices (corners) will be positioned. The other key feature that needs to be set out is the level. This refers to the height, or the depth, of the paving at any given point, and the gradients between various points, referred to as 'fall'.

### The Tools

On a large site, an automatic level might be used to set out the levels required at key points, but most DIY projects can be set out by using a spirit level, a straight-edged timber, a string line or a chalk line, and a pocket tape measure.

Not all that long ago, laser levels were limited to big construction projects, but they may now be found at many builders' merchant branches or large DIY stores for less than £50. However, with budget prices comes budget accuracy, and the small laser levels that rely on a bubble vial for levelling are notoriously prone to error. There are ways and means of establishing them and checking them to ensure that the laser line(s) generated are genuinely 'flat' or level, but they rely on the use of other tools. There are also self-levelling laser levels. These are simply established in position – placed on top of a pack of bricks or fixed to a tripod or some other immovable fixture – and the laser-generating apparatus inside is gravitationally adjusted to ensure that it is perfectly flat and/or plumb. These are far more accurate than the bubble vial models, but you will pay more for such a degree of accuracy.

As mentioned earlier when we looked at spirit levels, the longer the level, the more accurate it tends to be, so that a laser level based on a 200mm torpedo spirit level will not be as accurate as one based on a 600mm unit. While many professional contractors now have good quality laser levels as an essential component of their kit, for DIY-ers working on a project where falls are tight and accuracy is critical, it may be better to hire an accurate laser level for a weekend while the level set-out is completed.

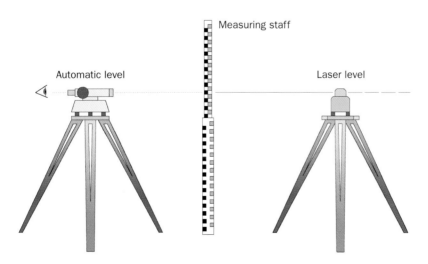

*Automatic and laser levels.*

## The Principles

All setting out of levels with these tools works on the same principle: a flat or level line is created across the site and the height above or the depth below that line is measured to determine and establish the required falls. Laser levels send out a line or a plane of laser light; automatic levels rely on what is, in effect, a small rotatable telescope that is set to look along a flat plane, while the spirit level is capable only of defining what is level and what is not.

On sites that have in-built natural gradients, it is often simply a matter of following the lie of the ground and of sending the water off over the edges on to a garden or into a gully, paying heed that the paving level remains at least 150mm below any damp-proof course (DPC), except for those few areas where level access to a doorway is required. However, many sites and projects will be flat or flattish and so falls and gradients need to be established to ensure adequate drainage. This is particularly true of patios, which tend to be relatively small and built at the back of a house in a fairly flat area.

As with the setting out of edge lines, the best way to demonstrate how the levels for a pavement on a flat site might be established is to work through an example. We shall return to the idealized driveway and describe the setting out of the falls and gradients that will be sufficient to ensure that no water stands on the paving when it is complete, a problem known as 'ponding'.

Happily, and as luck would have it, our driveway is almost exactly flat.

## Planning the Drainage

Although the site of the proposed driveway is flat, to all intents and purposes, there is a drainage system in place, although it may not be as extensive as that required for this project (see diagram on page 107). There is a simple hopper and trap drainage point located on each front corner of the house, along with one on the right-hand side of the porch, collecting rain-water from the porch roof. There is a further hopper and trap located on the corner of the garage block.

A circular inspection chamber (IC01) located approximately 1m from the north corner of the house has three inlets: one each from the drain points to house and garage, and a third that is known to link

to the rear yard area of the property. This IC is linked to a second, IC02, located about 2m from the porch, which also has inlets from the hopper adjacent to the porch and from the hopper at the south corner of the house.

The outlet of this IC runs east to a rectangular manhole (MH01) which is close to the proposed location of the gate pillars and which also has an inlet from the northern side that is thought to link to an old stable block, not shown on this plan.

Preliminary investigation of the hoppers reveals that they are set at a level 150mm below the DPC, the requisite minimum to comply with the Building Regulations, but there is no opportunity to reduce the level of the hoppers, since they sit directly upon the traps beneath.

A drainage scheme (see diagram on page 108) is drawn up that will achieve adequate falls, but additional drainage will need to be installed. The planned driveway has been divided into sections, with each being drained to a specific drainage point. Imaginary summit lines are drawn on the plan.

The level of the garage floor is 150mm lower than the DPC of the house and is therefore at the same level as all of the hoppers, the floor to the porch, and the public footpath. This level has been declared to have a value of 10.000m and all other surface levels will be established relative to this.

A new gully will be installed at G1. This will be connected into the drainage system via a new 300mm access chamber, AC01, fitted to the existing line. The surface level of G1 will need to be low enough to ensure an adequate fall to carry surface water from the furthest point in the section drained by G1 sector, which is marked as d1. The surface level at d1 needs to be slightly lower than that of the porch floor, so it is decided to establish it 15mm below, which is 9.985m. The summit line d1–d2 will be flat, so d2 will also have a surface level of 9.985m.

The distance between d1 and the gully at G1 is approximately 5.7m. A minimum fall of 1:60 is deemed to be adequate for a residential driveway; 1:60 is 0.0167, which is equivalent to near enough 17mm per linear metre, so:

$$5.7m @ 17 = 96.9mm,$$

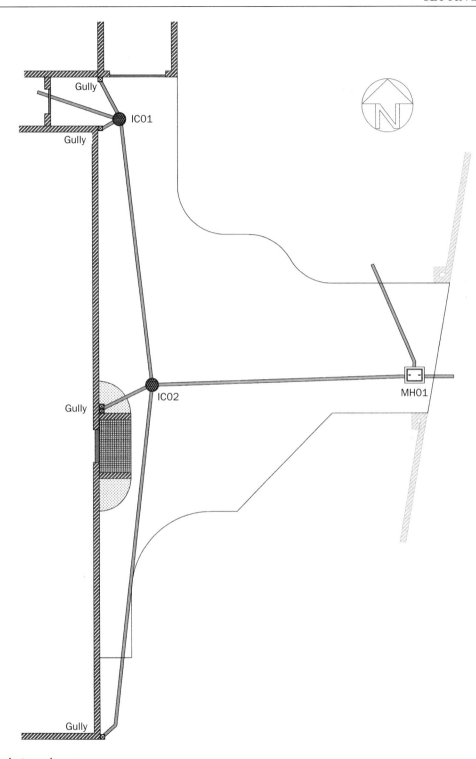

Gully

Gully

IC01

Gully

IC02

MH01

Gully

N

*Existing drainage layout.*

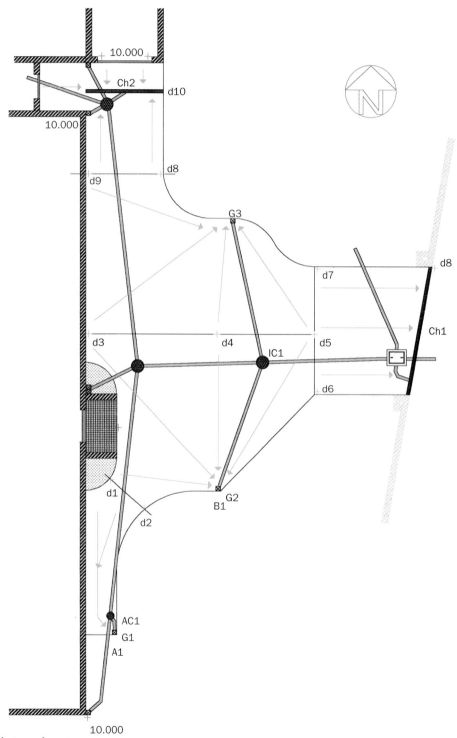

*Proposed drainage layout.*

and therefore G1 needs to be approximately 97mm lower than d1:

$$9.985 - 0.097 = 9.888m.$$

A second new gully will be installed at G2 and connected to a new inspection chamber (IC03) fitted on the existing drainage line (see diagram on page 110). As with the previous calculation, the surface level of the gully must be low enough to ensure an adequate fall from the furthest point, which is labelled d3 and which will have a surface level 15mm below floor level (that is, 9.985) and is approximately 7.6m from G2:

$$7.6m \times 0.017 = 0.129 \approx 130mm;$$
$$9.985 - 0.130 = 9.855m;$$

therefore, gully G2 will be established at a level of 9.855m.

At the threshold of the driveway, a new linear channel drain, Ch1, will be installed between the two planned gateposts and this will be connected into the existing manhole, MH01. The level of this drain must tie in with the public footpath, which is 10.000m. Fall needs to be generated from the summit line d6–d7, with d7 being the point farthest from the proposed linear channel:

$$d7 \text{ to channel Ch1} = 4.3m @ 17mm = 73.1mm,$$

so d7 needs to be at least 73mm higher than d8; we shall round up to 75mm, hence:

$$d7 = 10.000 + 0.075 = 10.075mm.$$

To create a slightly cambered profile, d6 will be established at the same level as d7 and the summit line linking the two will be established with its centre (d5) lifted by 25mm, that is at a level of 10.100.

Checking back to the channel at Ch1,

$$d6 \text{ to Ch1} = 3500mm \text{ with } 75mm \text{ fall;}$$

recalling that gradient = rise ÷ run:

$$75 \div 3500 = 0.0214 = 1:47.$$

It is also possible to check the fall from d6 back to G2, a distance of 5000mm, as follows:

$$10.075 - 9.855 = 180mm;$$
$$\text{thus } 180mm \div 5000mm = 0.036$$
$$= 1:28, \text{ which is ample.}$$

The summit line connecting points d3, d4 and d5 will be established as a 'flat bone', which means a straight line between d3 and d5. This has a length of 8800mm and a fall from d5 back towards d3 of 115mm (10.100 – 9.985), which means that there is a shade over 13mm of fall per linear metre, much less than is required to meet the minimum fall, but, as this is a summit line and not a fall line, it does not matter.

d4 is 3750mm from d5, therefore its level should be:

$$3.75 \times 13mm = 49mm \text{ approximately;}$$

we round off to 50mm, and so the level at d4 must be:

$$10.100 - 0.050 = 10.050m.$$

A gully will also be established at G3 and linked into IC03. We have already determined the surface levels at points d3, d4, d5 and d7, of which d3, at 7000mm, is the most distant. Once again using the 1:60 or 17mm per linear metre as a minimum fall:

$$7 \times 17mm = 119mm;$$

therefore the gully at G3 will need to be:

$$9.985 - 120 = 9.865mm.$$

Checking back for points d4, d5 and d7 gives:

$$d4 \text{ to G3} = (10.050 - 9.865) = 185mm;$$
$$185 \div 4200 = 0.044 \approx 1:23,$$
$$\text{and: d5 to G2} = (10.100 - 9.865) = 235mm;$$
$$235 \div 5200 = 0.045 \approx 1:22,$$
$$\text{and: d7 to G2} = (10.075 - 9.865) = 210mm;$$
$$210 \div 3500 = 0.06 \approx 1:17,$$

which is all well and good.

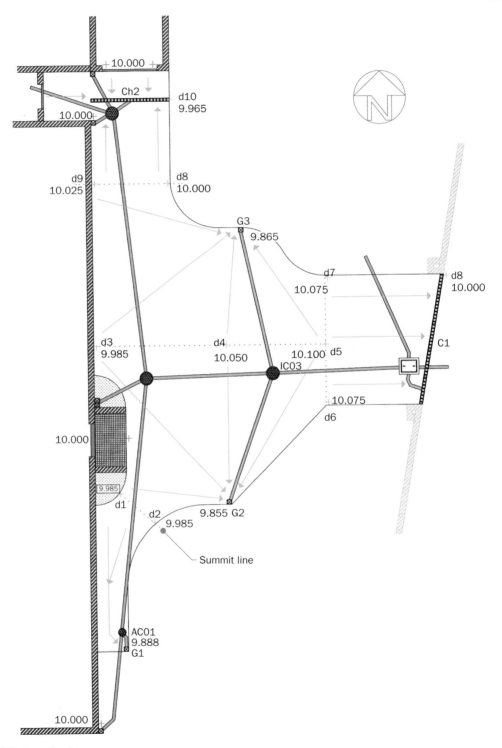

*Proposed drainage levels.*

The final drainage installation will be a 3m length of linear channel drain (Ch2) 1m away from the garage. This is being installed because each weekend cars are washed in this area, so installing a linear channel drain and connecting it to the IC already in place via an existing 'spare' inlet seems a sensible option.

The summit line runs between d8 and d9. The level at d9 will be elevated by 25mm to ensure that there is no backfall from the gully on the corner, the level of which cannot be altered. Although this means that the paving at this point will be less than the recommended 150mm below the DPC, it is a minor infringement over just a few metres, and so should be acceptable.

The distance between the proposed channel, Ch2, and the summit line is 3m. Using the 17mm fall minimum used previously, this means that the level of the channel has to be:

$$3 \times 17mm = 51mm \text{ below the level of point d9.}$$

We will increase this to 60mm to ensure adequate drainage for the car washing suds:

$$10.025 - 0.060 = 9.965m;$$

this level can be checked with d9 and the floor level of the garage;

$$\begin{aligned} d9 \text{ to } Ch2 &= 60 \div 3000 \\ &= 0.02 \\ &= 1{:}50; \end{aligned}$$

$$\begin{aligned} \text{garage floor to } Ch2 &= (10.000 - 9.965) \\ &= 35 \div 1000mm \\ &= 0.035mm \\ &= 1{:}29. \end{aligned}$$

We can also check that there is an adequate fall from the edge of the paving against the side gate and the channel:

$$\begin{aligned} \text{gate to } Ch2 &= (10.000 - 9.965) \\ &= 35mm \div 1800mm \\ &= 0.019 \\ &= 1{:}51. \end{aligned}$$

The final point to establish is d8. Setting this point to be level with its counterpart at d9 would be excessive and generate a considerable fall from d8 back to the gully at G3:

$$\begin{aligned} 10.025 - 9.865 &= 160mm \div 3200mm \\ &= 0.05 \\ &= 1{:}20, \end{aligned}$$

far more than is necessary. So we shall establish a level at d8 that ensures the minimal fall back to the channel at Ch2, which should then reduce the fall from d8 to G2 to a more suitable value.

From the earlier calculation of the level for d9 we know that the minimum fall required over the 3m distance to the channel is 51mm. Therefore, we shall establish d8 at only 35mm above the level of Ch2:

$$9.965 + 0.035 = 10.000,$$

and so the fall from d8 to G3 becomes:

$$\begin{aligned} 10.000 - 9.865 &= 135mm \div 3200 \\ &= 0.042 \\ &\approx 1{:}24; \end{aligned}$$

it is not much, but it will prevent the falls from looking excessive.

This exercise has shown how levels could be established. There are many other ways in which the levels could be determined: a minimum fall of 1:80, just 12.5mm of fall per linear metre, could have been used, although this is quite flat, even for residential paving. It could have been decided to install one central gully or length of linear channel drain close to the centre of the drive, and then dish everything towards that. There is often more than one way to drain a pavement and as long as it works, that is, as long as all the surface water is removed from the paving speedily and efficiently, it will be suitable.

We now have all the levels necessary to install the paving with adequate falls. All we need now is some way of transferring those levels from the drawing to the site, from theory to reality.

# ESTABLISHING LEVELS ON SITE

The three methods available would be to use an automatic level, a laser level, or a spirit level. Both the laser level and the automatic level work by producing a line, or, more accurately, a plane that is perfectly level and can be projected to any point on the site. The level or height of this plane is fixed and therefore any other point can be established by measuring down (or up) from this plane.

## Using an Automatic Level

An automatic level is set up at a convenient location from where all the other level points can be seen. This is likely to be somewhere close to IC02 near the porch. Once set up and checked that it is truly level, a staff is used to measure the distance between the line of sight (also known as the line of collimation) and the garage floor or the porch floor or the DPC level – in fact, any of the locations that were arbitrarily given a datum level of 10.000.

Looking through the lens of the automatic level towards the measuring staff, which has been placed on the porch floor, reveals that there is a level difference of 1.573m between the line of collimation and the floor level. Given that the floor level is said to be at 10.000, this means that the line of collimation must be at

$$10.000 + 1.573 = 11.573\text{m}.$$

The staff can be relocated to one of the points requiring a level to be established. We shall use the gully at G2 as an example: the lens of the automatic level is swung around so that it is lined up with the staff that has been positioned against a marker pin at G2; the required level is known to be 9.855m and we know the line of collimation is at a level of 11.573. Simple subtraction will give the required distance below the line of collimation for G2:

$$11.573 - 9.855 = 1.718\text{m}.$$

The staff can be moved up or down as necessary until the level 1.718 coincides precisely with the crosshairs within the lens. The bottom of the staff is now at the required level and may be marked on the pin.

## Using a Laser Level

The procedure when using a laser level is essentially the same. The laser is established at a convenient position from where it can generate a laser line to all the necessary points. The height of the laser line above a known point (the porch floor) can be measured by using a staff, or even a steel tape measure, and this distance is added to the datum level, exactly as was done for the line of collimation with the automatic level.

The laser can then be swung around to align with the points requiring their levels to be established and the same, simple subtraction used to determine the required level at any point on the site.

## Using a Spirit Level

The use of a spirit level is slightly less scientific, and somewhat less precise. The level plane, equivalent to the line of collimation or the laser line, can be transferred only by using a straight-edge or taut string line that has been checked for level with the aid of the spirit level. Any string line or straight-edge can only be as accurate as the spirit level used to check it, and it is inevitable that error will occur: the spirit level itself depends on a bubble vial for accuracy and on some spirit levels the vial may be slightly out. Secondly, the position of the bubble within the vial depends on visual judgement, and a millimetre to one side or the other can make a difference. Thirdly, the spirit level may not lie perfectly flat on the straight-edge or against the string line.

Transferring level points using a spirit level has limitations of accuracy. Transference across a distance of up to around 4m can be done with a single straight-edge, but the accuracy relies on the straight-edge itself being truly straight. With timber, 4m is about the limit beyond which the wood tends to flex too much. Steel or aluminium beams are less prone to flexing but there is a weight problem with steel, and then how many of us have a 6 or 7m length of aluminium I-beam at the back of the shed?

For the average DIY project a 4m length of $100 \times 50\text{mm}$ timber is about as much that can be expected. To transfer level points more than 4m would therefore require the process to be done in stages, with the use of intermediate level stations. Alternatively, a taut string line could be used. To rely

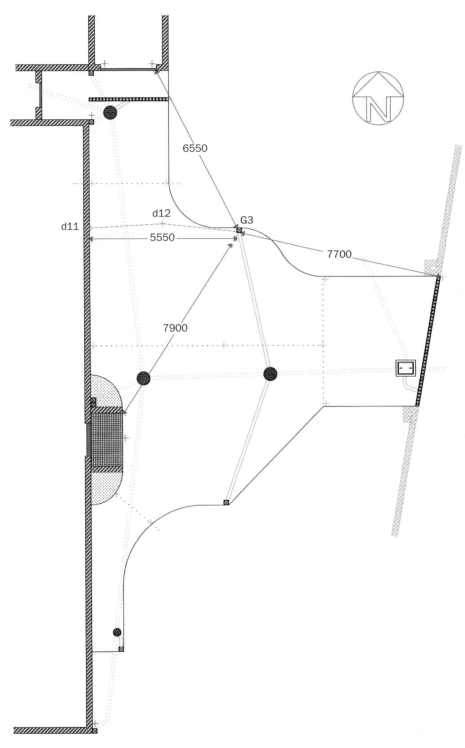

6550

d12

d11

G3

5550

7900

7700

*Establishing a level for gully G3.*

on a straight-edge and intermediate stations results in greater and greater inaccuracy the more stages are involved, while to use a taut string line is also limited because, no matter how taut the string is pulled, any distance of more than around 10m will result in some sag in the line.

*Level Transfer*

Referring back to our idealized driveway, to set a level for the gully at G3 by using just a spirit level presents a challenge (see diagram on page 113). If the porch floor level were to be transferred, the distance is around 8m. The garage is a little nearer, at only 6.6m, while the public footpath at the drive threshold is around 7.7m distant. The nearest point at which there is a known level is actually the front wall of the house, perpendicular to the gully at point d11, a mere 5.5m away. The level of the DPC, or of the paving level two courses down from DPC, could be transferred by using either a taut string line stretched from d11 to G3, or a 3m straight-edge with just one intermediate station.

Using the taut string line method, a marker pin would need to be established at d11, and one at, or close to, G3. Alternatively, a masonry nail could be hammered into the mortar joint at d11. The string needs to be fastened at d11 and stretched as tightly as possible to the pin at G3, where a half-hitch loop can be used to fasten it to the pin. The spirit level has to be aligned to the string line, and this is easiest to do by resting one end on the line where it is tied to the pin and then adjusting the position of the far end of the spirit level until its base is just touching the line. The height of the line fastened at G3 will need to be adjusted up or down until the level check indicates that it is precisely positioned, or as near as possible. The level of the half-hitch may then be marked on the pin – a piece of adhesive electrical tape is ideal for this task, although a pencil or marker pen will do just as well.

The straight-edge method requires a platform at d11 that is capable of supporting the straight-edge itself. This could be the masonry nail mentioned above, but more commonly it would be a paving block or broken piece of flagstone bedded on a shovelful of sand and tapped down until the top

---

### Fastening String Lines to Marker Pins

Reef or granny knots are rarely used when fastening string lines to marker pins since they may often result in permanent knots within the string itself, which, in turn, reduce its accuracy. Two simple hitch knots are preferred because they are temporary, easily loosened and do not result in any permanent knotting of the string.

The simple hitch, also referred to as a 'half-hitch', relies on tension created in the string line as it is stretched from one pin to another to 'trap'

a loop of the string against the pin itself, and so hold it firmly in place. A cow hitch is used at the start of a string line, where the end of the string is fastened to the first marker pin. It uses a double loop to fasten the line against the pin. Unlike a clove hitch, which involves a crossing-over of the string, the position or level of a cow hitch on the marker pin is easily adjusted up or down as required without needing to unfasten the knot completely.

LEFT: *The simple or half-hitch is used to fasten a string line to an intermediate or end-of-line pin.*

RIGHT: *The cow-hitch is used to anchor a string line to the first pin of a series.*

*A level can be transferred in stages: from a known point to an intermediate station …*

*… and then from the intermediate station to the required location.*

is flush with the required level. The block or flagstone needs to be reasonably firmly bedded as any movement of the platform could result in an error. One end of the straight-edge is seated on this first platform and aligned in the general direction of G3. A second, intermediate platform is then created at a position (d12) that will allow the free end of the straight-edge to be rested upon it, while the far end is still in position at d11. The level of this platform is adjusted by tapping down the block or flag piece until the spirit level reports that the straight-edge spanning platform d11 and platform d12 is level. The straight-edge is then moved so that it now spans from platform d12 across to the marker pin at G3, where it can be supported by hand or a further temporary platform could be created while the process is repeated, this time transferring the level from platform d12 to G3.

# CHAPTER 9

# Drainage

It has been known for at least a couple of thousand years that one of the key factors in creating a successful pavement is drainage, and the guiding rule has been to get any water off and away from the paving as quickly and as effectively as possible. It is only over the last decade or so that we have started looking at permeable or porous paving systems, where the surface water is directed into the pavement rather than away from it. However, for the typical residential patio or driveway, the more traditional approach of removing the water as quickly as possible is still the most commonly recommended plan.

We shall divide the subject into two parts: first, how the water is collected and channelled, and secondly, how it is disposed. Taking collection and channelling first, there are two forms of drainage that will be considered here – point drainage and linear drainage.

## POINT DRAINAGE

This is simply a drainage system in which the surface water is directed to a single point from where it is passed into the underground pipework. These 'single points' are usually gullies; there may be more than one of them on any project, but each is referred to as a 'point' and the paving in its vicinity is usually laid so that it slopes or 'falls' towards that particular gully.

### Gullies, Traps and Hoppers

In some parts, gullies are referred to as 'grids' or 'drains'. The grids are actually the gratings that

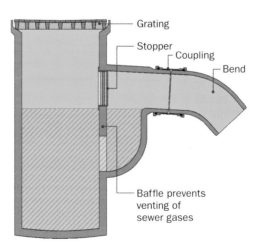

*Gullies are 'trapped' by means of a 'baffle' that blocks the escape of sewer gases.*

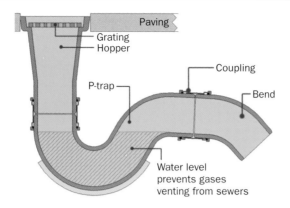

*A trap and hopper assembly relies on the U-bend feature of the trap to block the venting of sewer gases.*

cover the top of the gully and prevent larger pieces of litter and detritus from entering the system. Gratings come in all shapes and sizes, ranging from those with relatively narrow openings seen on most fittings for residential use, to the considerably wider openings seen on the gratings used on public highways.

Gullies are basically a 'pot' or a chamber into which surface water is directed before being discharged into the underground pipework system. A common feature of gullies is the 'trap', which is a sump or a low point within the structure that performs two important functions: it traps sediment and other heavy materials, including valuables, and prevents their being carried any further into the drainage system, and it also prevents gases and what are termed 'odours' escaping from the sewers and venting onto a patio or driveway. This is achieved by having some form of U-bend incorporated into the design of the gully. Sediment and other heavy materials cannot rise up the outlet side of the bend and enter the pipework, while gases of all descriptions cannot pass through the water retained by the bend and escape through the inlet.

Gullies are usually single piece units. The outlet pipe, known as the spigot, connects to the underground drainage pipe and a grating sits on the inlet opening. An alternative is the 'trap and hopper' arrangement, which separates the two essential components. It may not be immediately apparent just why it would be considered necessary to separate the trap and the spigot component from the inlet

*Clayware: a typical yard gully. (Hepworth)*

*Plasticware: a square hopper fitted to a P-trap.*

117

hopper, but the advantage this offers is that the hopper can be skewed to suit the site layout. Most gullies have the spigot in a fixed position, and, if the hopper part of the gully is fixed square to a building, then the spigot is also square to the building, which may complicate the sub-surface pipework or even make it impossible to connect up. With a trap and hopper arrangement, the trap and spigot element can be twisted to any angle required, and so the connection to the sub-surface pipework is simplified. Some more modern gullies now incorporate a twistable hopper inlet that does not need to be aligned square to the spigot, but many of the older gullies are fixed and offer less flexibility and freedom to the installer.

## LINEAR DRAINAGE

This refers to those systems that allow water to enter at any point along their length. The two most common forms are dished channels and channel drains. Both operate by collecting water from a larger area and delivering it to a disposal point.

### Dished Channels

A dished channel is simply a specially shaped paving unit with a concave profile that allows water to run along its surface. Such channels range in size from short, usually decorative or coloured units of pressed concrete or fired clay, up to 900mm or larger units of plain pressed concrete that are more commonly used for civil and commercial schemes. They are typically

laid to falls on a bed of concrete and may have the joints sealed or mortared to prevent water escaping. They normally deliver the water to a point drain, that is, a gully. It may seem nonsensical to use such a channel to send water into a gully when the water would simply run over the existing paved surface if the channel were absent, but these channels are a neater solution, ensuring that the water follows the path determined by the designer, which does not always coincide with the natural fall of a paved area, and they keep it all in a relatively narrow stream, rather than allowing it to spread out over a metre or so, as often happen with non-channelled layouts. The only real disadvantages are that the dished profile may be a hazard for bicycles, wheelchairs, prams and the like if the channel is in the centre of a pavement. For this reason, channels are often placed at the edge of a paved area, where traffic is less likely to be inconvenienced. Secondly, the water remains on the surface, so there is still a risk of splashing and ponding.

### Channel Drains

These fittings also use a dished channel, but the 'dish' is typically much deeper, usually U-shaped in profile rather than concave, and is enclosed by a grating. This counters both of the disadvantages described above: the surface profile is essentially flat and so presents no hazard to wheeled traffic, and the water is carried just below pavement level, out of sight and out of mind. However, these features bring their own

*A dished-channel unit manufactured in clay.*

*Many linear channel drains for residential use are manufactured from recycled plastic and offer a choice of grating. (Aco)*

minor disadvantages. The gratings, if not selected carefully, can be more of a hindrance than a dished channel would be, and the fact that the water-carrying channel is out of sight means that blockages and other problems may not be noticed until the drain overflows. However, despite these failings, these units are becoming increasingly popular as prices drop and availability spreads.

Channel drains are usually made from a polymer concrete (a high-strength, fine grained concrete that can be moulded to thicknesses of only 6mm or so) or from some form of plastic. The plastic models tend to be cheaper, especially those manufactured from re-cycled materials, but they are not as tough as the polymer concrete models. However, for virtually all patio and driveway projects, the plastic channels are perfectly suitable.

As with the dished channels, the channel drains are normally laid to a gradient on a bed of concrete, although short lengths (5m or less) may be laid flat, as any water falling into the channel will find its way to the outfall as its level rises. For commercial and civil projects, there are special units with an in-built fall. Some models feature a form of tongue-and-groove arrangement for linking successive units, although, in practice, any slight gaps between units soon fill up with sediment and effectively self-seal after a few weeks.

There are a variety of ways of connecting channel drains to the sub-surface pipework. Some models include a special 'outfall' unit that may include a trap, but the most common method for residential installations is to rely on a base outfall or end outfall connector that links up to a trap unit and from there into the drainage system proper.

Channel drains and their gratings are 'graded' according to what sort of loads they can be expected to withstand when laid correctly. It is a simple grading system, Classes A to F, where Class A is used for those units suitable only for light pedestrian and cycle usage, and Class F is the sort of thing that would be used on an airport runway. Class B is the type normally used for residential driveway use. Separate classes are awarded to the channel and to the grating, since it is possible that, for instance, a Class D channel might be used with a Class B grating. In such cases the overall class of the combined

*The linear channel on completion of the driveway.*

channel and grating is always that of the weakest component, so, in the example given above, the installed channel and grating would qualify as a Class B installation.

Once installed, the grating is the only visible part of a channel drain, and the selection of a suitable grating can have quite an affect on the finished appearance of a pavement. Some of the budget drainage channels include a tinny, galvanized steel grating, with a nasty tendency to distort as soon as a child's cycle runs across it. Many of the better plastic models include a sturdy plastic grating with relatively narrow openings that present less of a hazard to women's heeled shoes, are less liable to bend or break, and simply clip into place. Although black and grey are the most common colours for these gratings, others are available, albeit at a premium, and there are also gratings specifically manufactured to be heel-proof or suitable for use with particularly narrow wheels.

## DRAINAGE SYSTEMS

Having looked at the most popular methods of collecting surface water, we now move on to how the water is delivered to the sub-surface system and how it is disposed.

### Surface, Foul and Combined Systems

There are, presently, three main types of drain system used. The surface water system (SW) collects all the water from driveways, patios, pathways, balconies, roofs, and other surfaces open to the weather. This water is relatively 'clean' and is often discharged into a local watercourse, a soakaway or a modern SUDS scheme (sustainable urban drainage system – a sort of twenty-first-century soakaway that is considered later). The foul water system (FW) collects all the dirty water (kitchen waste, toilets, bathrooms and the like) and takes it to a treatment plant where it is cleaned up before being discharged into the environment. The third type is a combined system, where the same network of pipes collects both SW and FW.

Cleaning up water is expensive, so sewer design tries to ensure that any clean water (SW) is kept completely separate from the FW and returned to the environment as soon as possible. This is why many post-war properties have two drainage systems – one for the SW and one for the FW. Combined systems were used in the earlier part of the twentieth century, and in exceptional situations after the Second World War, but they are used only when there is no other option, since they result in huge volumes of perfectly clean water being put through the treatment process, adding costs for local ratepayers and reducing the capacity of the treatment plant.

When draining a typical driveway or patio, any water that cannot be returned direct to the environment by whatever means (it may be possible to have a soakaway or SUDS installed specifically to deal with this surface water) should be directed into an SW system or a combined system, if that is what is present. Only as a last resort should surface water be delivered into an FW system, and, when this happens, any connections made to the FW system must be trapped to prevent sewer gases from escaping.

Thus when considering how a pavement should be drained, the options, in order of preference are

1. Local soakaway or SUDS
2. SW system
3. Combined system
4. FW system

## DRAINAGE MATERIALS

There are two principal types of underground 'drain' pipes used in Britain and Ireland: those made from clay and those made from uPVC plastic. Existing installations may use cast iron pipework or pipes made from the dreaded asbestos or pitch-fibre, but most residential drainage dating from before 1980 will be 100mm or 4in internal diameter (ID) clay pipes of one form or another, while most modern installations (from 1990 onwards) tend to be 110mm ID uPVC. There was a transitional period from 1980 to 1990, and some modern developments still use clayware (as it is called) in preference to plasticware. There are pros and cons associated with each type, but, in general, for residential DIY projects, plasticware is slightly easier to work with.

## DRAINAGE COMPONENTS

Sub-surface drainage systems generally consist of a range of pipes and fittings. The pipes are simply long, straight tubes that are joined together or cut to size as required, while the fittings form the remainder: the bends, the junctions, the traps, the hoppers, and the couplings. Some systems have a coupling element incorporated within the pipe or fitting, while others rely on separate couplings that are fitted to each component as required. Again, there are pros and cons to each, with little to choose between the two when one is considering their suitability for residential projects.

Pipes, fittings and couplings are joined together to create a network that collects water from various sources and delivers it to a main sewer, which is simply a much bigger and more robust version of the residential drainage. Pipework of 100 or 110mm ID usually relies on push-fit couplings. There is no 'glue' or welding involved because the couplings are fitted with a special rubber-like ring that slides over the pipe and forms a watertight seal. Before joining together the several components required to link, say, a gully to the rest of the drainage system, the couplings are usually lubricated with a soap-like gel to make it easier to slide them into position.

### Pipes

Pipes are typically laid on a bed of small gravel, sometimes known as pea-gravel, and then covered with the same material to a depth of around 100mm. The gravel acts as a cushion, protecting the pipe, evening out any load from above or around, and acting as a

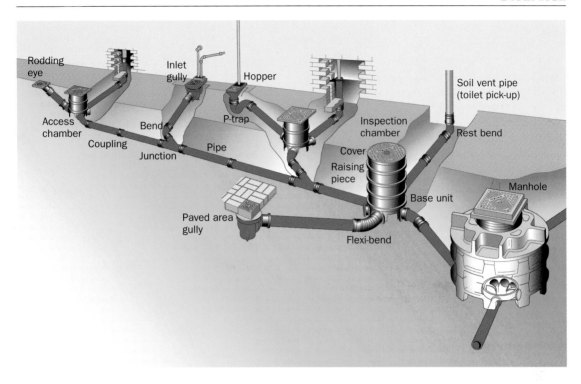

*Popular drainage components for residential projects. (Hepworth)*

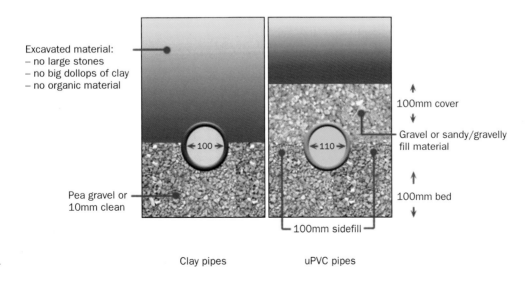

Excavated material:
– no large stones
– no big dollops of clay
– no organic material

Pea gravel or
10mm clean

100mm cover

Gravel or sandy/gravelly
fill material

100mm bed

100mm sidefill

Clay pipes

uPVC pipes

*Cross-section of pipes laid in a trench, showing bedding and sidefill and cover requirements.*

sort of early warning system for any future excavations. Where pipes and associated fittings are at an exceptionally shallow depth, they may need to be laid on and encased within concrete. This is particularly true of plasticware drains that are less than 600mm beneath a driveway; in such situations, it is worth considering a switch to clayware, which is less prone to deformation (being squashed by the loads imposed on it). For patios and pathways, all drainage should have at least 300mm of cover.

The pipes themselves have to be laid with 'fall' so that the water runs the right way, that is, away from the collection point and towards the disposal point, wherever that might be. For residential drainage, the fall on any SW pipe or fitting should be not less than 1:100 (1cm per metre), as an absolute minimum. The required falls for FW or combined systems are steeper – 1:80 as an absolute minimum, with 1:40 preferred.

### Access and Inspection Chambers

An essential feature of any drainage system is some provision to gain access for inspection and maintenance purposes. The Building Regulations governing drainage require that a set of drain rods or a CCTV camera should be able to reach every part of a drainage system. This is achieved in a number of ways, but the three most relevant for residential projects are rodding eyes, access chambers (ACs) and inspection chambers (ICs).

Rodding eyes are simple extensions to a line of

drainage, bringing a pipe up to surface level where a special cover can be removed and drain rods inserted. Access chambers tend to be part of the system rather than an extension to it; they are often relatively small diameter 'chambers' (225 or 300mm in diameter) and are usually at less than 1m depth. They have a cover fitted at ground or pavement level which can be removed to gain access. Inspection chambers are simply a bigger version of access chambers; typically, an IC is a rectangular or circular chamber having an internal dimension of 450mm or more, through which an open section of the pipeline flows.

## CONNECTING DRAINAGE

### Installing a New Chamber

When new surface drainage needs to be installed, the simplest method of connecting the new pipework and fittings to the existing system is via an access chamber or an inspection chamber. Many post-1980 installations feature circular ICs or ACs with preformed bases which often include a 'spare' inlet. In such cases, it is a relatively simple matter to excavate around the IC, expose the spare inlet, remove the 'cap' that prevents mud and earth from entering the IC while the inlet is not being used, and then connect up the new pipework.

Older systems may have ICs built from brick or concrete, and to make a new connection into these can be much more of a challenge. In some cases it is actually easier to install a new, preformed AC or IC on to the existing drain line than to break through all the brickwork and/or concrete.

The task of inserting a new AC or IC into an existing pipeline is fairly straightforward, although it does usually involve some strenuous digging and fiddling with uncooperative components. In essence, a length of pipe is sliced out of the existing line and replaced with the AC or IC, which will have at least one spare inlet that can be used to accommodate the new drainage.

*Locating and Excavating*

The first task is to locate the existing pipeline and determine the best position for the new chamber. The easiest method of locating pipelines is by aligning other ACs and ICs. Drains run in straight

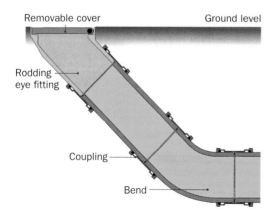

*Cross-section through typical rodding eye.*

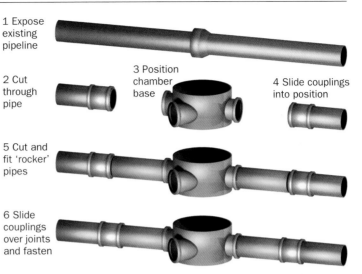

1 Expose existing pipeline

2 Cut through pipe

3 Position chamber base

4 Slide couplings into position

5 Cut and fit 'rocker' pipes

6 Slide couplings over joints and fasten

*ABOVE: Access chamber base unit with three inlets and one outlet.*

*RIGHT: Key stages when installing a new access chamber.*

lines (or at least they do in theory), so if there are two chamber covers on a driveway, it is a safe bet that the pipe linking the two of them lies in a dead straight line between them. Where such obvious clues are not available, it may be necessary to dig down at an existing drainage fitting and trace back the pipework to a point suitable for the insertion of the new connection.

It is important to give yourself plenty of working space when excavating for a new chamber. Even though the IC to be fitted may have an external diameter of only 500mm, digging a hole only 700mm in diameter will result in much struggling and cursing, whereas one of 1000 or 1200mm diameter will make the task ahead much easier. The excavation needs to go down to 100 to 150mm below the bottom of the existing pipe. This is because the AC or IC to be fitted has a base of its own that requires some space, and it is best to seat the new chamber base on a few shovelfuls of semi-dry or moist mix concrete so that it can be set to the exact level required.

*Removing the Old Pipe*

Once the pipe is exposed, a length of it needs to be cut out in order that the new AC or IC can be inserted. The new base unit will require three short lengths of pipe, known as 'rocker pipes' to be fitted – one to the upstream inlet, another to the downstream

outlet, and the third to the inlet that will take the new connection. These rocker pipes should extend from the base unit by at least 150mm. The base and the rocker pipes can be assembled at ground level, and when complete the whole assembly should be accurately measured and this measure transferred to the existing pipeline, allowing 10–15mm extra for play. The pipe is best cut out using a power cut-off saw or a large angle grinder, which can be hired in, if necessary. Before making the cut, it is a good idea to put a stopper into the pipeline at the nearest point upstream of the working area. This will help to keep the working area dry and clean until the new chamber base is installed. However, if it is not possible to fit a stopper upstream, the cuts can be made and a stopper fitted into the upstream end of the pipeline once the section is cut out and removed.

Do not make any cuts immediately next to an integral collar (as found on many older clayware pipes), as this will make it impossible to fit the adjustable couplings required to connect the old pipework to the new chamber base. If a collar or a coupling is present near the planned new chamber, it is best that the chamber be centred on the joint, so that two 'good ends' of pipe are available upstream and downstream. Saw straight through the pipe, keeping the cut as vertical as is possible, referred to as 'square to the barrel of the pipe'. If the cuts are angled too severely (say more than 10 degrees from plumb),

123

there could be problems in achieving a watertight seal with the new chamber base. In some situations, the only option is to make an angled cut because of restrictions of space. In such cases, once the section of pipe has been removed, a new cut should be made to square up the angled cut, or the angled cut should be trimmed to render it as near square as is feasible. Try positioning the new chamber base assembly into the gap left by the removed pipe section to ensure a good fit. If a short make-up piece is required, this can be measured now, attached to the base unit assembly, and the whole checked once again.

When the pipe section has been removed, and it has been checked that the base assembly will actually fit, the upstream and downstream ends of the pipeline should be exposed by digging all the way around them with a small spade or trowel. Approximately 150–200mm of pipe should be exposed all around and cleaned of any clay, sand, mud or other material. A sliding, adjustable coupling is then placed over both of the cut ends and pushed back so that the whole of the coupling sits on the pipe barrel.

*Fitting the New Base Unit*

The concrete for the base unit may now be put into position. An ST1-equivalent concrete is perfectly adequate for this task, and it should be spread and levelled over the entire base excavation. The base unit assembly is then offered into position and tapped down into the concrete bed with the aid of a small rubber mallet until the rocker pipes are aligned with the exposed ends of the existing pipeline. Check how level the base unit is using a spirit level. If the inlet and the outlet are assumed to be at the 12 o'clock and the 6 o'clock position, then the spirit level should span from 3 to 9 o'clock. If the base is leaning one way or the other, the whole chamber will reflect this when it is installed. Once you are satisfied that the base is level and firmly bedded, the adjustable couplings previously fitted to the pipe ends can be slid into position on the rocker pipes, so that the coupling bridges both the pipe end and the rocker pipe. It may then be tightened up as recommended by the coupling manufacturer (this normally requires a spanner or a screwdriver, depending on the exact type of coupling being used).

It should be mentioned that, if preferred, the adjustable couplings may be loosely fitted on to the rocker pipes and slid on to the pipe ends when in position. In practice, manipulating the base unit and its rocker pipes is enough of a hassle without the worry that the adjustable couplings have not slid loose nor fallen off. That is the hard work done. If the upstream pipeline had been stoppered-up, the stopper should be released and any backed-up water should flow through the new chamber base without hindrance.

The base unit can be secured in place with additional concrete and the new pipework, linking on to that third rocker pipe, can now be installed. The raising piece(s) can be slotted into the chamber base, the frame and the cover fitted (even if it is only temporary) and the hole backfilled.

## Backfilling Drainage

When backfilling excavations for drainage it is important that the pipework is protected, that any material put back into the hole or trench is suitable, and that it is properly compacted. It is always a good idea to ensure that all pipes and fittings have a covering of gravel, sand or other small-sized granular material that will cushion and protect them from damage. When it comes to the general backfill, there is not much point in using sloppy or soft clay, or organic materials such as paper, timber or old leaves.

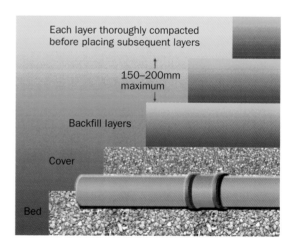

*Backfilling should be done in layers.*

The soft clay will never firm up sufficiently and any organic material will decompose, resulting in a void that will lead to settlement. The material dug from the trench during the original excavation is usually suitable, as long as it is inert and reasonably granular in nature. Do not backfill using big lumps of clay, or boulders that will leave voids; what is needed is a material that will leave no voids, so if in doubt, leave it out. Make good any shortfall in backfill material with gravel, all-in ballast, or a suitable sub-base material such as crusher run or Type 1.

To achieve good compaction it is essential that the backfilling is done in layers, as a series of 'lifts', where 150–200mm of material is placed into the trench or the hole and thoroughly compacted before adding the next layer or lift. This is particularly important under any areas that are to be paved. If a little settlement occurs beneath a garden area or the lawn, it is not the end of the world, but if such settlement occurred beneath a driveway or a patio it will need attention to correct the problem and eliminate any ponding. A narrow vibrating plate compactor, rammer, or a trench compactor are ideal tools for this, but if none is available, a punnel or heavy hammer may be used to ram down the material in each layer before progressing to subsequent layers.

The backfill should be finished off level with the top of the prepared sub-grade, that is, at the underside of a sub-base layer (if one is being used) or the laying course material for lightweight pavements. As with sub-base preparation, it is critically important that the 'surface' of the backfill has a tight finish, with no open voids or holes down which bedding or sub-base material could disappear, since this could result in settlement of the finished paving.

## SOAKAWAYS

We have previously considered drainage that connects into a larger network of pipes for disposal at some point off-site, possibly a local river or even a treatment works. However, one of the major changes in drainage planning over the past decade has been the realization that, by sending all this reasonably clean surface water down miles of pipes and straight into our national watercourses, we are actually exacerbating the problem we have with flooding, and reducing the opportunity for natural aquifers (water-carrying rock strata) to refill themselves. Water that once fell on an open field or meadow now lands on the roof or driveway of a home or office, gets diverted into an underground pipe, and is eventually spewed out possibly miles away from where it originally landed, into a river or stream that is already bursting its banks. The land itself has a massive storage capacity and it acts as a natural 'balancing reservoir' for the environment, storing water when it first arrives from our grey and leaden skies and gradually releasing it to the local waterways when levels dictate. It is not perfect, and it does not mean an end to flooding, but it is quite effective at reducing flood risk.

So, the old-fashioned technology of a soakaway has been rejuvenated, updated and brought into the twenty-first century. It is now incumbent on new

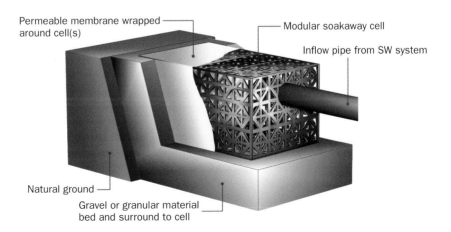

Permeable membrane wrapped around cell(s)

Modular soakaway cell

Inflow pipe from SW system

Natural ground

Gravel or granular material bed and surround to cell

*Diagram of basic modular soakaway installation.*

*Soakaway cells in position. Note the permeable geotextile that will wrap the cells and the inlet pipe feeding into one of them.*

For new residential pavements, it may be tempting to install drainage that simply discharges into the existing SW system, but it is well worth considering installing a simple SUDS soakaway beneath the garden, or even beneath the pavement itself. They are easy to install and you would actually be doing a favour to the environment. Some water companies are even offering rebates on water bills for properties disposing of all their surface water to soakaways and other SUDS installations.

Obviously, there is rather more to it than merely digging a hole in the lawn, dropping in a soakaway cell, and backfilling with whatever came out of the hole in the first place. For a soakaway to function it has to be capable of allowing water to soak away, so it must be positioned within a permeable sub-stratum, above the water table, and at least 5m from any building. The hole excavated should be large enough to hold the modular cells, allowing for 100mm of gravel or free-draining material at the base, the modular unit itself and then at least 500mm of 'cover'. The gravel or free-draining base is there primarily to help with the levelling of the modular unit(s) and to ensure that they are firm and secure. Once the gravel is in place and levelled out, the hole can be lined with a permeable geotextile before placing the cell(s). The connecting pipework is fed into one of the cells and then the geotextile is neatly wrapped around the whole assembly, enveloping all four sides and the top.

The gap around the cell(s) should be backfilled with gravel, or some suitable free-draining material, and then the whole lot can be buried with excavated material, making sure that it is compacted in a series of individual layers, as explained previously. There may be some settlement around and above the soakaway over the first few months as groundwater washes through so it is worth 'crowning' the backfill, building it up slightly higher than the surrounding ground. As and when the backfill settles, the surplus earth can be levelled out as required to suit the site.

The modern cell soakaways are incredibly strong, so much so that it is actually possible to install a soakaway beneath a driveway, if no other option is available. The manufacturers and suppliers are able to provide the necessary detailed installation guidance.

developments that all surface water, that is, the water from roofs, paths and roads, is sent not to the local SW system, but delivered to a SUDS installation, from where it can be gradually discharged into the underlying sub-soils and rocks. Unlike older soakaways, modern SUDS installations tend to be huge affairs, covering tens or hundreds of square metres and having the capacity to hold hundreds of thousands of litres of water at any time; and, unlike older soakaways, they are no longer simple holes in the ground filled with any old rubble, half-bricks and broken flagstones that happened to be lying around. The modern SUDS soakaway uses a modular cell component that can be thought of as a plastic, reinforced box used to create as large an underground void as is possible without compromising its load-carrying ability. The modular nature of these cells allows many units to be linked together, creating a soakaway large enough to cope with the anticipated influx of stormwater. So, for example, a single-property SUDS soakaway might use two cells linked together, while a soakaway for a cluster of homes might use twenty, fifty or a hundred of the cells, all linked into one massive unit. These larger assemblies are usually referred to as 'attenuation' systems, because they have a relatively large storage capacity and are designed to retain water during storm periods, allowing the level within the cells to rise as more water flows in, and then to release it to the ground over a much longer period of time.

**CHAPTER 10**

# Laying Block Paving

## INTRODUCTION

When it comes to laying patios and driveways, the type of surfacing to be laid usually dictates the most appropriate laying method. However, there are some materials that can be laid in a variety of ways. For instance, block paving is usually laid as 'flexible paving', on a bed of sand, with secured edges and sand joints, but sometimes it is laid as 'rigid paving', on a bed of mortar with mortar joints. The most suitable method is determined by considering the exact type of material (rigid block paving is normally constructed using clay pavers), the planned usage (rigid is a more decorative finish), the skill level of the laying operative (rigid laying requires more highly skilled labour and takes longer to complete), and the required appearance, of course.

This chapter and the following one examine the most popular laying methods for modular paving, namely, the individual bedding method and the screed bed method. These core techniques may be used with paths, patios and terraces, with flagstones, and with setts and cubes. The text in this chapter follows the step-by-step construction of a real driveway, paved with a tumbled block paver, using the screeded bed laying method, as this is to be a flexible pavement. It is tempting to think that a residential driveway would be 'stronger' if it were laid by using the rigid method, with lots of concrete and cement holding everything in place, but flexible construction is the most popular construction method for block paving and is actually extremely strong and resilient, despite the use of cement-free sand for both bedding and jointing. It is also simpler,

*Rigid laid pavers usually have mortared joints ...*

*... while flexible laying often uses sand-filled joints.*

faster, and, when done properly, it is perfectly suitable for all residential projects. Later chapters will consider rigid and individual bedding that build on the techniques introduced below.

## GETTING STARTED – SETTING OUT AND EXCAVATION

All projects start with the preparation, which involves clearing the site, setting out and excavating to formation level. Site clearance may simply be a matter of getting rid of obstructions such as old patio furniture, washing lines and garden features. It may involve digging out shrubs or perennials and transplanting them elsewhere in the garden, or it might be a matter of lifting old paving, lawn, weeds or whatever else is there to give the blank canvas needed for setting out.

### Site Clearance

On our block paving project, the old flags on the driveway need to be lifted and carted away, as do the edgings, and there are plants to be excavated and moved to the back garden. The existing flags are quite large – 900 ¥ 600 ¥ 50mm, what used to be called 'three by twos' in the days of imperial measurements – and they are awkward to handle. Each flag of

*The project driveway at the start of the works.*

this size weighs around 65kg (10 stone 3lb) – so to lift and dump them is no mean feat.

There are three options:

- Lift, clean and sell as second-hand
- Smash up and load into a skip
- Smash up and feed into a crusher.

To keep the flags intact and look for a buyer would be the preferred solution, but care is needed when handling and stacking them. A sign attached to the stack offering them to passers-by may attract a sale, as might a small advertisement in a local paper. In some parts, salvage dealers will 'take them off your hands', but they are unlikely to cross your palm with silver.

Smashing up the flags renders them easier to handle. A few blows from a sledgehammer should break them into manageable chunks, but eye protection such as goggles or safety spectacles must be worn, and do not break the flags while near any windows. Once broken, the pieces can be loaded into a wheelbarrow and dumped into a skip. It is not a good idea to pile as much broken concrete into a skip as is physically possible because the wagon may not be able to lift a skip fully laden with broken concrete; 50mm concrete flags have a density of approximately 120kg per m² and 100m² of flags weigh around 12 tonnes, more than should be loaded into a typical builder's skip. On average, load no more than 80m² of flags to a skip, and even that may be excessive; check the weight limits with the skip company before filling, since having to partly empty a skip is not only hard work, it is soul-destroying when you have already strained your back and blistered your hands loading the material in the first place.

There is often a temptation to use broken flags and old bricks as some form of sub-base. Throwing rubble into a skip is not cheap and some people have the mistaken belief that half flags will make an excellent foundation for their new drive or patio. Sadly, it is not so, and to use half-flags, old bricks and assorted rubble as a sub-base will actually reduce the service life of a patio or driveway, unless they have been broken into pieces smaller than 100mm and then capped with a 'proper' sub-base of granular material. Given the high cost of disposal, it is now becoming

*Flags must be stacked safely by leaning them against a flat stack.*

*Old flags can be broken up to make handling easier and safer.*

economically viable to hire a portable crusher machine to break up old concrete and concrete products. These machines are fed broken flags, kerbs, bricks, concrete and other hard, inert materials (so *no* tarmac), and a few seconds later they spit out a granular material of reasonable quality that is acceptable sub-base material. The cost of hiring these machines varies, but, as a guide, when there is more than 100m² of concrete or flagstones to dispose of, they can be cost-effective when compared with the cost of hiring a couple of skips and buying the sub-base material.

## Setting Out

Once the site has been cleared the setting out work can be done. The theory and mathematics behind setting out were considered earlier; all that needs to be borne in mind is that an allowance should be made for spread and working room, and having string lines stretched hither and thither can, at this stage, be more a hindrance than a help. Spray-paint is a convenient alternative, as are sand-lines: they indicate the extent (but not the depth) of the excavation.

## Excavation

Although this may be estimated during the planning stage, the actual depth of excavation can be finally decided only once the ground is open and the condition of the sub-grade inspected and assessed.

For the block-paved driveway, the minimum depth of excavation below the finished paving level (FPL) is calculated as:

block thickness + bed depth + sub-base depth,

which, for this project, is calculated as:

60mm + 40mm + 150mm = 250mm.

It may be that, at a depth of 250mm below FPL, the sub-grade is still soft or unreliable and further excavation is required. It is essential that any suspect material, any soft spots, anything that does not seem firm, sound and reliable should be excavated and removed. It is much easier to excavate an extra barrowful of

*Densities and Skip-load Capacity for Regular Materials*

| Type of material | Density (t/m³) | Quantity per skip (m³) |
|---|---|---|
| topsoil | 1.45 | 5.1 |
| clay | 1.80 | 4.4 |
| cinders/gravel | 1.60 | 4.8 |
| concrete | 2.25 | 4.0 |
| bitmac | 2.10 | 4.2 |
| old bricks | 2.10 | 4.2 |

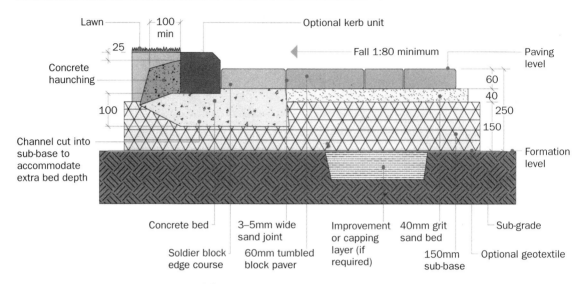

*Cross-section for residential block paved driveway.*

suspect ground at this stage than to wait for any movement or settlement in six months' time and have to lift and reconstruct parts of the paving.

For smaller areas, hand digging is the only viable option. A good spade, backed-up with a pick or mattock, can break the ground, load it into a barrow, and then it is a matter of getting it into skips or losing it elsewhere on site. Increasingly, mini-diggers are used since they dramatically increase productivity, can cope with relatively tough ground and have no trouble in loading the excavated spoil into skips. Care is required to avoid damaging sub-surface service pipes, cables and drains, but most of these should be well below formation level. Some hand digging

around surface fittings, such as gullies, access covers and cable inlets, will be required since it is too risky to rely on a mechanical digger so close to fragile or dangerous fittings.

In general, excavated spoil is carted off-site and builders' skips are the most commonly used disposal method. Larger projects may require wagons, but these can be loaded only by full-sized excavators and are supplied on a daily basis rather than on a 'per load' basis and so tend to be used only for projects that will generate 40m³ or more of spoil.

## Build-up

### Capping Layers

Following excavation, the sub-grade at formation level may not be perfectly even and will not necessarily reflect the final profile of the paving. Some areas may have been excavated to 250mm while other spots needed additional excavation to remove soft material, so the exposed sub-grade may resemble a cratered moonscape or a bombsite. Sub-base material or other suitable granular material can be used as an improvement layer to level out any unevenness, filling low spots and generally aiming to produce a reasonably level surface for the sub-base proper. Material used as an improvement layer should be

| *Cartaway Capacities* | |
|---|---|
| **Container** | **Capacity (m³)** |
| wheelbarrow | 0.1 |
| midi skip | 2.5 |
| builder's skip | 4.5 |
| drop body | 6.0 |
| 6-wheel wagon | 7.0 |
| 8-wheel wagon | 9.0 |

placed in depths of not more than 150mm and thoroughly compacted before any further material is added. On completion, the improved surface should be reasonably level, indicating the required profile of the finished paving as far as possible, although further 'adjustment' is possible when installing the sub-base. The next consideration is whether or not to use a geotextile.

## Membranes

Geotextile membranes are primarily used to 'improve' the load-bearing capacity of a pavement. This is achieved by preventing sub-base material from commingling or disappearing into the sub-grade beneath and also by helping to spread the loads imposed. They have a secondary role as a weed barrier and are particularly useful on sites subject to pernicious weeds such as Mares' Tails and Japanese Knotweed since they keep the emerging shoots out of the pavement structure (although this is achieved by deflecting shoots so that they emerge at the sides or edges of the membrane).

Geotextiles are rolled out to cover the sub-grade and temporarily weighted down with half-barrows of sub-base material to prevent the wind from whipping them away. Joints between adjacent pieces of geo-textile should be overlapped by 300–400mm. Sometimes, joints are taped with geotextile tape or 50mm packaging tape, although this is not strictly necessary.

## Sub-base

As mentioned previously, a sub-base is not always necessary for paths, terraces and patios, but should always be present beneath driveways or other pavements expected to carry vehicular traffic. Most sub-bases will be 100–200mm in depth – 150mm is a typical figure. Where a sub-base greater than 225mm in depth is required, it should be constructed in separate 'lifts', with each lift not exceeding 225mm and being thoroughly compacted before the placing of any covering layer(s). In practice, the compaction equipment used for residential paving is not really suitable for compacting any depth of sub-base material greater than 150mm, so even a 225mm sub-base might be constructed in two lifts – one of, say, 125mm with a 100mm second lift. A 300mm sub-base could be constructed as two lifts of 150mm, while a 400mm sub-base could be constructed as four lifts of 100mm each or three lifts of 135mm.

The sub-base material is moved into position either by wheelbarrow or by being dozed along by the blade or bucket of an excavator. It is levelled out by using a spade/shovel and a rake until it is reasonably even, with no humps or hollows and roughly at the required level.

It is not possible to give a table of values stating exactly what depth of uncompacted granular material is required to give a particular compacted thickness. Just how much any material will compact depends on the type of material being compacted, the moisture

*A geotextile separation membrane installed between laying course and sub-base is intended to prevent loss of sand to any voids in the sub-base. For most installations the separation membrane is better placed between sub-base and sub-grade, as explained in the text. (TDP)*

*Compaction of Sub-base Material*

| Uncompacted Depth (mm) | Approx Depth Following Complete Compaction (mm) |
|---|---|
| 70 | 50 |
| 100 | 75 |
| 130 | 100 |
| 165 | 125 |
| 200 | 150 |
| 235 | 175 |
| 265 | 200 |
| 300 | 225 |

*The sub-base material is levelled out by using spades and then 'fine-tuned' with a rake to achieve the desired level.*

*A vibrating plate compactor is used to consolidate the sub-base material.*

*The accuracy of the sub-base may be checked by 'dipping' and measuring down from a taut string line.*

content, and the size, type and power of the compacting equipment. However, we can generalize, and it is a fair assumption that approximately 200mm depth of uncompacted Type 1 material at optimal moisture content will compact to around 150mm thickness when hammered down with a minimum of eight passes of the vibrating plate compactor. Other approximate values are given in the table on page 131.

Over the years, experiments have been carried out on several sub-base materials at varying thickness with various types of compacting equipment. For those interested in the findings, these are summarized in *Specification for Highway Works*, available from The Stationery Office or construction bookshops; all that it is necessary to know, as far as residential paving is concerned, is that sub-bases up to 150mm thick need at least eight passes of a good vibrating plate compactor to achieve that state of happiness

referred to as 'compaction to refusal' – the point where the material simply will not compact further, no matter how many times the plate passes over it.

Once the sub-base has been compacted its level needs to be checked. This is best done by stretching a string line between two known level points or by using a straight-edge and then 'dipping' with the aid of a small tape measure, to determine the depth between the finished paving level and the level of the compacted sub-base. Normally, the sub-base is dipped every metre or so to check its level.

The usual sub-base tolerance is ±15mm. On our block paving project we are using 60mm blocks and a 40mm laying course, so the sub-base should be exactly 100mm below the FPL, but we could accept levels in the range of 85–115mm below the FPL. Any high spots, that is, any areas of compacted sub-base with a surface level less than 85mm below the FPL, need to be scraped down and compacted again.

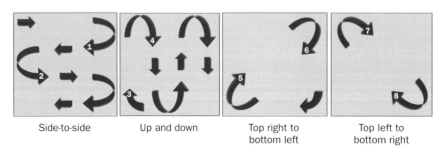

Side-to-side      Up and down      Top right to bottom left      Top left to bottom right

*Eight passes are the minimum required to compact a sub-base.*

Similarly, any low spots need to have additional material spread, levelled and compacted over them to achieve the required level. This is known as 'regulating' the sub-base.

One other consideration regarding the sub-base is 'tightness'. As the sub-base comprises granular material containing particles up to 40mm in size, it is possible that some patches of the compacted surface will consist mostly of these larger particles and will have an 'open' or honeycombed appearance, while other patches will mostly consist of smaller particles, the fines, and will therefore have a 'tight' or close-textured appearance. It is this tight finish that is desirable. This is because the laying course material could trickle down into the voids or interstices of an open-textured sub-base, which would cause a drop in level of the laying course and the subsequent settlement of the paving. Where an open texture is apparent, it should be tightened-up by applying additional fines or grit sand, brushing it into the surface, compacting again with the vibrating plate, if necessary, or even loosening the compacted material, adding fines and/or grit sand and then compacting once more to a (it is hoped) tight finish.

As with the efforts made in regulating the finished level of the sub-base to bring it within the acceptable tolerance limits, effort spent in ensuring a close-textured finish will reduce the risk of settlement.

## KERBS AND EDGE COURSES

For certain types of paving, edge courses are purely decorative or might be omitted completely, but for block paving some form of robust edge restraint is essential. They are a key component of the pavement and must be properly constructed if the paving is to last more than a season or two. Where the blockwork abuts a permanently fixed object, such as the walls of a house or a garage floor slab, fixed edge courses are not essential, but where the pavement lies against an open area, such as a garden or a lawn, these so-called 'free edges' must be fixed and immovable, otherwise the paving will move with them. If preferred, kerbs or decorative edgings may be used as restraining edges; rather than relying on block edge courses. This is perfectly acceptable, provided that the kerbs or edgings are permanently fixed in position. This is the

*On the left, the sub-base is too open due to lack of fines, while on the right it is an ideal 'tight' finish.*

key requirement for *all* edges on a block pavement: they must not be capable of moving, shifting, giving way or collapsing. They lock in position the rest of the paving and their integrity is vital to the integrity of the completed pavement. Whether that immovability results from the edge courses being against the side of the house, or whether it is because there is a line of blocks laid on and haunched with a generous portion of good-quality concrete does not really matter. As long as they are firm and fixed they will do.

On our project driveway there are three differing forms of restraining edge. Against the house itself, will be a 'soldier course' of blocks laid as headers tight against the wall; adjacent to hedges and garden areas the client requires the same simple, soldier edge course, but it will need to be held in place with a concrete haunch; and on lawn edges the client wants a low kerb, which must also be laid on and haunched with concrete. However, to complicate matters, two different forms of kerb are needed. On the left-hand side the lawn is higher than the planned driveway and so a kerb with 25–50mm of upstand will be used; on the right-hand side the lawn is 50–150mm lower than the planned driveway and so the kerb will form a step down to the lawn and will be edged with another block to form a mowing strip.

This gives a total of four types of edge construction; these are shown in the illustration on page 135.

### Laying on Concrete

The contractor has decided that the soldier edge courses against the house wall will be laid on a lean-mix concrete bed. It may seem that to use a concrete bed in such a position is unnecessary since the blocks should not be able to go anywhere because of the wall. However, the contractor knows from experience that they will be using these edge blocks as a screed guide (see later) and that blocks laid on sand alone are easily dislodged or accidentally forced down into the sand bed during the screeding process. By using a lean-mix concrete the blocks will have a firm base, holding them fast when the screed board passes over, and the bond formed between the concrete and the blocks will prevent their being dislodged until the cutting-in is completed.

ST1 concrete will be used for laying all the edge courses. This is a good compromise and avoids the nonsense of using different strengths of concrete for different parts of the construction. ST1 concrete (roughly 1:3:6) is strong enough to support the blocks and the kerbs, and, if properly compacted, makes a good haunching. The blocks and the kerbs will be laid on a semi-dry or moist mix of concrete since this suits the laying gang's methods. There is less 'floating' of adjacent blocks or kerbs and the units tend to stay put once tapped down into position.

A taut string line is established as a guide to line and level. For straight runs the kerbs and/or blocks will just touch the line without disturbing it, but, for arcs and radii, the edge units will be some distance from the string line and so a spirit level will be needed to check the accuracy of level. The accuracy of alignment may be checked by measuring from the origin of the arc, assuming that it is available, or by relying on 'eyeing in' for those arcs and radii where measurement from the origin is not possible.

The concrete is lined out where required. The bed should be at least 50mm deep (100mm deep for free edges and kerbs), so it may be necessary to scoop out some of the sub-base material, creating a shallow channel for the concrete bed and haunch. For the relatively shallow soldier course blocks, concrete is lined out roughly level with the taut string line, while for the deeper kerbs the level of the concrete will be reduced appropriately. The fresh concrete is trampled down and levelled off using a brick trowel, spade or float, to a bed level that will leave the unit approximately 10–15mm high. The unit is offered on to the prepared bed and tapped down to level with a rubber mallet.

Any surplus bedding on the inside edge of the block and/or kerb is removed so that the bedding flares out only slightly. This is done to ensure that a full bed of grit sand can be laid and that the blocks

*Concrete Coverage for Various Types of Edging and Kerb*

| Kerb/Edging | Bed and Haunch | Concrete per Linear Metre | Linear Metres per m³ of Concrete |
|---|---|---|---|
| 100mm-wide blocks | 50mm bed only | 0.006–0.008 | 100–133 |
| | 50mm bed/75mm haunch | 0.013–0.017 | 43–58 |
| | 100mm bed only | 0.011–0.015 | 50–67 |
| | 100mm bed/100mm haunch | 0.022–0.029 | 25–34 |
| 200mm-wide blocks | 50mm bed only | 0.011–0.015 | 50–67 |
| | 50mm bed/75mm haunch | 0.015–0.019 | 30–40 |
| | 100mm bed only | 0.023–0.030 | 25–33 |
| | 100mm bed/100mm haunch | 0.033–0.044 | 17–23 |
| Small kerb (125 × 125mm) | 100mm bed/100mm haunch | 0.033–0.044 | 17–22 |
| Large kerb (200 × 125mm) | 100mm bed/100mm haunch | 0.034–0.045 | 14–19 |
| Large kerb (200 × 125mm) laid flat (200mm wide) | 100mm bed/100mm haunch | 0.042–0.057 | 13–18 |
| Edging kerb (50 × 150mm) | 100mm bed/100mm haunch | 0.028–0.037 | 20–27 |

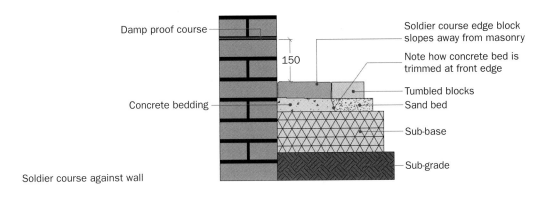

Soldier course against wall

Damp proof course

Concrete bedding

150

Soldier course edge block slopes away from masonry

Note how concrete bed is trimmed at front edge

Tumbled blocks

Sand bed

Sub-base

Sub-grade

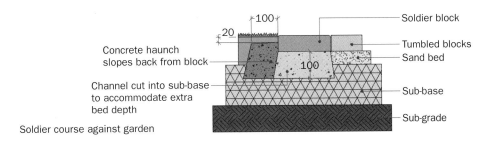

Soldier course against garden

Concrete haunch slopes back from block

Channel cut into sub-base to accommodate extra bed depth

100

20

100

Soldier block

Tumbled blocks

Sand bed

Sub-base

Sub-grade

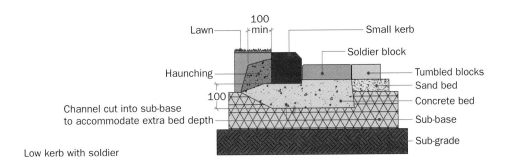

Low kerb with soldier

Lawn

Haunching

Channel cut into sub-base to accommodate extra bed depth

100 min

100

Small kerb

Soldier block

Tumbled blocks

Sand bed

Concrete bed

Sub-base

Sub-grade

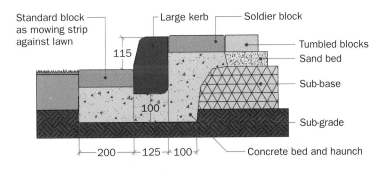

High kerb retainer

Standard block as mowing strip against lawn

115

100

200   125   100

Large kerb

Soldier block

Tumbled blocks

Sand bed

Sub-base

Sub-grade

Concrete bed and haunch

*The four types of edge course used on the project driveway.*

*The edge course blocks are laid on a bed of moist concrete, using a taut string line as a level guide.*

*Surplus concrete should be removed from the 'front' or inside face of the edge course blocks to avoid problems with differential settlement.*

that will eventually lie in this position are not bedded partly on concrete and partly on sand, which can cause problems when it comes to final compaction.

Mortared joints with edge courses are not standard practice, although they may be used on arcs and radii. A better finish is achieved when blocks presenting a joint width greater than around 10mm are 'taper cut' to close up the joint. Similarly, kerbs are not normally mortar jointed, but on layouts involving several arcs, the finished look can be improved if all kerb joints are mortar pointed, rather than display gaping joints of varying width on the curves and tight butt joints on the straights.

## Checking Alignment

Once the laying is complete, the alignment should be checked by observation from a variety of positions. Look along straight lines from both ends to ensure that they are actually straight. Walk around arcs and curves to ensure that they are 'sweet' and present a regular, even swing. Just because the taut string lines or the tape measure says that such-and-such a block needs to be in such-and-such a position does not mean that this will look right to the eye. It is often possible to construct an edge course or kerbline that is mathematically accurate but simply looks wrong. The mathematics do not matter to observers of the finished driveway: all they will be concerned with is

whether it looks right or not, so trust your eye for any final adjustments.

Blocks laid tight against walls may need to be eased outwards slightly to present a pleasing line along the front edge. Any gap between block and brickwork can be filled with jointing sand to keep the blocks in line. Blocks laid to curves may need to be nudged back a few millimetres, only to find that they now look even worse, so nudge them back and try moving the adjacent blocks. Keep walking around, visually checking the alignment, nudging, tapping and adjusting until you are satisfied – when it looks right, it probably is right.

## Haunching

Once you are happy with the alignment, the haunching can be done. The same concrete mix could be used for the haunching or it could be 'wetted up' by adding a small quantity of water, just enough to make it 'sticky'. Haunching is placed at the rear of the edge units, so that it is at least 75mm in width – any less will not provide sufficient strength. It should be patted down with the boot or a trowel to compact it, and then 'polished' by running over it with a trowel or a spade to give a smooth finish. The top of the haunching should be approximately 25mm lower than the top of the block or kerb, and should be sloped 'backwards' so that water

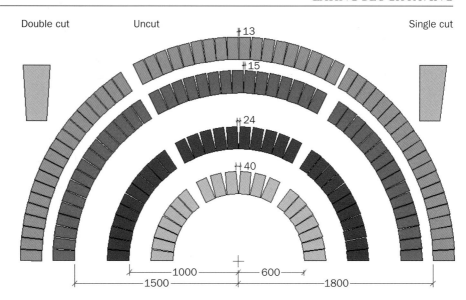

Double cut       Uncut       Single cut

13
15
24
40

1000 — 600
1500 — 1800

*Single-edge and double-edge taper cuts.*

cannot collect on top. As shown in the large kerb cross-section, where paving will be laid above haunching it should be kept low enough to accommodate the paving and some bedding, and similarly with haunching covered with turf – although grass will grow in 25mm of soil, it is better to keep down the level by an additional 10–25mm to provide a little extra growing medium for turf. However, the structural integrity of the paving must always take precedence over the needs of a little grass – there is not much point in having a lush and verdant sward if the edge of the driveway is collapsing.

Finally, once haunching is complete, check the alignment one last time. It is quite possible that placing the haunching has disturbed the previously perfect positioning, and any irregularities can be nudged back into line while the concrete is still plastic. Any concrete spilled on to kerbs or blocks should be washed off immediately since it will stain. Use a soft brush and clean water to clean off any staining.

Ideally, edge courses and kerbs are left overnight to allow the concrete to harden but this is not always possible, and, if work must continue immediately, care is needed to ensure that edge courses and kerbs are not dislodged nor accidentally knocked. Any

*The concrete haunching should be smoothed off and kept 20–30mm below the top of the blocks/kerbs.*

*For the low kerb edge, the kerb is laid first and aligned before the edge course blocks are laid.*

*By laying the blocks and kerbs as one structure, the strength of this retaining edge is enhanced.*

misalignment should be corrected immediately, since the concrete haunching will be getting stiffer and harder by the minute.

## THE LAYING COURSE AND THE NOBLE ART OF SCREEDING

On the project drive the edges are now complete. The next task is to get the laying course material into place and to prepare it for block. This is achieved by the technique known as screeding. In essence, this involves placing the grit sand, lightly compacting it, and then scraping off the excess to create a surface that is level and smooth and begging to be covered by the paving.

There are three 'approved' methods for preparing a screeded bed, but one is more or less useless, the second gives highly variable results, and so we shall concentrate on the third, which has worked well for my paving company over the last thirty years and more. It is known as the 'pre-compaction method'. (For those keen to learn more about the other methods, some detail is given in BS7533, Part 3 and a number of Interpave publications.)

### The Right Stuff

First, the sand has to be ferried into position, and it is critically important that the correct sand is used. As discussed previously, grit sand is the preferred material. Stone dust, whin dust or grit may be used if that is what is available locally, but building sand, soft sand, masonry sand and the like are simply not acceptable; do not waste your time and money by hoping that the uselessness of such sands is over-stated; it is not, and if you decide to use them your driveway *will* settle and rut and start to collect water and look botched.

So, the correct sand is on site and it has been properly managed by being sheeted over or retained in the delivery bags to prevent it drying out or becoming too wet. It is barrowed into position, tipped out and spread using a shovel, rake and/or a lute to give a fairly even coverage of sand that, in its uncompacted state, covers the edge blocks to roughly two-thirds of their depth. There is no need to be excessively accurate with the levelling – as long as it is reasonably even, so that the bed is of a fairly uniform depth, it will be fine. And there is no need to cover the entire area with sand in one operation: spread sand only over the area that you are confident of getting paved in that session; loose sand will only attract the local cat population to visit the site overnight and leave little surprises for you the following day.

### Compaction

Next, bring in the plate compactor and compact the levelled-out sand with one or two passes, no more. Compaction to refusal is not required; in fact, to do so would be a problem. What is needed is 'almost compaction to refusal' so that there is still a little give in the sand. This is because the blocks (or flags, setts or whatever) have some natural variation

*The laying course sand is tipped in piles ...*

*... and then levelled out with a rake.*

*Once levelled out, the sand is compacted with one or two passes of the plate compactor.*

in thickness; 60mm blocks are manufactured to a thickness tolerance of ±3mm, so, in theory, one block could be 57mm thick while its neighbour is 63mm thick, a difference of 6mm. In actuality, most of the top manufacturers achieve far better tolerances, but there is always some variation and therefore the give in the sand bed is essential to accommodate this variation and give a smooth, level, finished surface. So, two passes of the plate and then remove it from the area. It will not be needed again until the blocks – or flags, setts, etc. – are laid.

## Screeding: Art or Science?

As mentioned earlier, screeding basically involves scraping off surplus sand to create a smooth, even bed for the paving. So we shall need something with which to do the scraping and something to guide the level of the scraping, making sure that just enough of the compacted sand is removed, but not too much.

The tool that does the scraping is the screed board. This might be a length of straight and true timber or a length of aluminium strut, or it might even be one of the expensive screeding tools manufactured specifically for this purpose. The only requirements are that it should be light enough to handle comfortably, strong enough not to bend or distort when dragging a weight of sand and, most importantly, it must be straight. If a bowed length of timber is used the result will be a bowed screed bed. The length of the board is not critical. Obviously, if the drive or patio is exactly 2.5m wide, a 3.6m screed board is not going to be of much use unless it is sawn down. Most paving contractors have a selection of screed boards, ranging from short 600mm lengths up to about 4m; boards longer than 4m are unwieldy and difficult to manage, although screed boards of this length, and longer, are used, particularly on commercial projects where they are dragged along by a tractor. For most residential paving, a length of timber around 3m length is ideal; 100 × 19mm board, 75 × 50mm structural timber, 100 × 50mm timber – whatever is comfortable. There is no right or wrong, as long as the conditions listed above are observed.

Next come the screed guides. There are two methods of guiding a screed board as it removes excess sand; one uses the edge courses as level guides and relies on having a 'notched end' to the board; the second relies on extra items of equipment known as 'screed rails'.

### Notched End Screeding

This is the preferred method since it requires no extra kit and ensures that the levels created are directly matched to the edge courses. It is as simple as cutting a notch in one end of the timber and then allowing the shoulder of the notch to 'ride' over the edge courses. The notch is usually 30–80mm long – as long as there is sufficient shoulder to sit comfortably on the edge course, but not so much that it interferes with the use of the board, the length is not critical. However, the depth of the notch is highly critical; the depth determines just how much of the surplus sand will be scraped off. It may seem obvious – laying a 60mm block requires a 60mm-deep notch, doesn't it? If only it were that simple!

Recall that the sand bed is not compacted to refusal, that there is still some give in it. Also allow for the variation in block thickness which needs to be accommodated. In fact, there is no simple formula that gives a notch depth for any given thickness of block; a 60mm block might require a notch depth of, say, 53mm – when laid, the blocks will be around 7mm high and this will be reduced once the plate compactor has been used. However, on a subsequent job, it may be found that the 53mm notch depth

| Notched Screed Board Depths | |
|---|---|
| **Thickness of Pavers (mm)** | **Depth of Screed Board (mm)** |
| 30 | 23–26 |
| 35 | 27–30 |
| 40 | 32–35 |
| 45 | 37–39 |
| 50 | 42–44 |
| 60 | 51–54 |
| 65 | 56–59 |
| 70 | 60–64 |
| 80 | 70–74 |
| 100 | 90–94 |
| (Assuming bed depth of 40mm) | |

*Where a pavement has parallel edges, a screeder board can be custom-made to fit the gap with a notch at each end.*

leaves blocks too high or even too low. The problem is not so much the screed board but the sand used for the laying course. There are many variables that affect the extent to which the sand will compact once the blocks are laid. Obviously, the type of sand has some bearing – a sand from the Pennine hills will compact differently to one dredged from an Essex gravel pit, or hewn from a rocky crag off the west cost of Scotland. How much compaction can be expected from a particular sand is not the only variable though, we also need to factor in the moisture content, the total bed thickness, the weight and force of the compaction equipment, the texture of the sub-base, the weather conditions and also quite probably what day of the week it is.

We can generalize: there are typical values that usually work, most of the time, all other things being equal, but the only way to be certain is trial and error. Use the values given in the table on page 139 to prepare a suitable screed board, lay a couple of square metres or so of the paving, run the plate compactor over them and see how much they go down. If it is too much and the blocks are lower than the edge courses, saw off the existing notched end and cut a new one that is not quite as deep; if the blocks are left too high, shave off another couple of millimetres or whatever is needed to produce the correct level.

On some projects the pavement width will be constant and the edges parallel. If this width is less than, say, 2m, it is possible to cut notches on both ends of a measured screed board so that it fits neatly

between the edge courses, with 20–50mm of play to prevent it becoming jammed tight. On other projects, the width will be more than the maximum feasible with a single screed board, or the width will vary, or the edges will not be parallel, and so some other method of guiding the screed board must be created. This brings us to screed rails.

*Using Screed Rails*
The principle behind screed rails is similar to that of the notched-end technique, except that instead of having a shoulder ride over the edge course blocks, the base of the screed board rides over a rail set into the laying course.

Many sorts of item are used as screed rails. Some contractors use lengths of timber or steel reinforcement bars, some have sets of interchangeable aluminium sections that link, while some use garden canes. If it works and produces a screeded bed at the correct level, there is no way in which it can be condemned. Many contractors find that tubular steel of the type used as electrical cable conduit is probably the best material: it is tough enough to resist bending, stiff enough not to sag under its own weight, while its hollow construction renders it reasonably lightweight and, best of all, it is cheap and endlessly reusable.

The screed rail is bedded into the laying course so that its top is at the exact level required for the screeded bed. This level is determined in the same way as that for cutting the notched end, described above, that is, a mixture of experience and guesswork.

The positioning of the screed rails needs to be considered. On the driveway project, the paving at the drive threshold is 6m wide. There is an edge course or kerbline to each side that can carry a notched end, so a screed rail positioned centrally will permit the screed board to be dragged through to screed off the left-hand side of the drive, with the notched end on the edge course, while the right-hand end rides over the screed rail. Then, it can be flipped over so that its left-hand end is now on the screed rail while the notch rests on the right-hand side edge course. The completed screed will have been completed as two halves, with the screed rail positioned at the meeting point.

A groove slightly wider than the screed rail is

formed in the sand bed by using a brick hammer or chisel. The rail section is placed into the groove and tapped down until it is firm along its entire length. The level of the top of the screed rail needs to be checked: this is best done using the 'dipping' technique and a taut string line, checking the screed rail level with one of the paving blocks, packing up with extra sand or knocking down with additional taps from the rubber mallet, as required. If a transverse line is being used, reposition the string line so that the rail can be dipped every metre or so. With a longitudinal string line that runs along the line of the screed rail, use a block to dip continuously along the rail, or at least every metre or so. The importance of accuracy when levelling up the screed rail cannot be overstated: whatever levels are set at this stage will determine the finished level of the paving, so check, double check and then check again.

*The Screeding Process*

This can be done by one person, but for screed boards longer than 2m it is much better if two persons work the board, each being responsible for monitoring the level at their end. Nestle the bottom of the board into the sand until the shoulder of the notched end is resting on the edge course and/or its base rests atop the screed rail. It is usually most comfortable, and most effective, if the screeding operatives are kneeling in the sand and gently but firmly draw the board towards their bodies, using a slight sawing motion if necessary to remove surplus sand. As sand builds up in front of the screed board,

it can be removed with a shovel or trowel so that the weight of sand being dragged is kept to a minimum. It may take two or more passes of the board to reduce the sand to the required level. Any areas found to be low should be built up with additional sand that is compacted by using the flat of a trowel or a small punnel, and then these areas should be screeded again to create the correct level.

Eventually the screed board should have passed over the entire area, scraping off the sand excess and leaving behind a wonderfully smooth and level bed for the blocks. *Do not walk on this bed.* In theory, this screeded bed is now equally compacted throughout and any footprints will result in a greater degree of compaction in that spot than in the surrounding areas, which could be enough to produce a slight depression in the finished surface. The golden rule is that, once a screeded bed has been prepared, nothing and no one should be allowed to stand on it.

*Final Touches*

The only exception is when it comes to removing the screed rail. This is usually withdrawn before block laying commences and the channel left by its removal has to be filled, compacted and screeded off to match the areas to either side. This can be done only by walking on part of the area already screeded to level: carefully withdraw the screed rail from the bed and move it clear of the working area; using a shovelful of surplus sand, top up the channel, trampling it in, making sure that it remains slightly proud of the screeded surface, and gradually progress into the

*Use a paving block to check the level of a screed rail.*

*Once the screed rail is in place, one half of the bed can be screeded to level …*

*… and then the other half.*

*The screed rails are lifted out and removed …*

*… and then the bed can be finished off as required.*

*An alternative strategy is to lay blocks up to the screed rail before removing it.*

screeded area. Keep any disturbance to a minimum, aiming to work in a narrow line around 200mm wide. Once the channel has been filled, use a float or a short screed board to remove the excess and gradually work your way backwards, out of the screeded area, levelling and titivating the bed as you go.

An alternative method is to screed off just one 'width', leave the screed rails in place and then pave the prepared area, almost up to the line of the screed rail. Once an area has been covered with blocks the screed rail may be lifted out and the channel left in its wake filled with sand. The rail is repositioned so that the adjacent width can be screeded, using the 'just laid' blocks to guide one end of the screed board while the repositioned rails guide the other.

There is nothing wrong with this method and it certainly does away with the need to titivate any screed rail channel, but some contractors find that they prefer to prepare a larger screeded area in advance and work through, covering a substantial width (or length) of pavement in each laying session without repeated breaks while another screed width is prepared. However, to screed from uncompacted blocks is less than ideal: the blocks at the leading edge (the 'open' edge where the next blocks would be laid) can be knocked out of position and, because the leading edge is usually 'stepped', the screed board has to be constantly repositioned, jiggling in and out and around the blocks on the edge. Further, this edge has to be titivated to remove any surplus sand left there by the screeding process. This is often overcome by removing the blocks at the leading edge and then running a short screed board or a float over

the surface to tidy any tailings and ensure a smooth, level transition between the two separately screeded areas.

At this stage, there should be a perfectly formed, screeded bed of laying-course sand. Any apparent ridges should be smoothed out with a float or a short screed board. Any surplus sand left against the edge courses should be carefully removed with a trowel or a float. Walk around the prepared area (without walking on it) to check its accuracy from as many angles as possible. The profile of this screeded bed is, to all intents and purposes, the profile of the finished paving, so spend whatever time is needed to ensure that it is perfect. And then we can get down to the block laying.

## BLOCK LAYING

This is what it is all about – getting the finished surface into place. This is actually one of the speediest tasks, and definitely the most satisfying.

There is another one of those 'golden rules' to be borne in mind when it comes to block laying: *never work from the screeded bed.* Balance on the edge course or work from outside the pavement area to get the first few blocks into position, and then work from on top of the blocks that have been laid. As more blocks are laid, the area from which the laying can be carried out will increase, but that first square metre or so can be awkward and time-consuming. However difficult it may seem, do not stand on the screeded bed and lay blocks from there – the screed will be disturbed and the blocks that eventually cover the area trampled in the process will be at the wrong level.

The laying pattern should be decided in advance, taking into account the type of block, the layout of the site and the personal taste of the customer paying for it all. Herringbone patterns give the greatest degree of interlock but can be tricky to establish; stretcher courses are simpler, but do not give as much interlock and look overly simplistic for some types of block, while working well with the tumbled types; basketweave patterns are best used only for patios, pathways and other areas with no vehicular traffic; other, block-specific patterns are possible but there are so many of them that it is not practicable to document them all in the space available.

Whichever pattern is selected, it has to be aligned. A stretcher course should run across the direction of traffic, but, if a layout is not square to the house, it might be best to align the courses to run parallel to the front of the property. For herringbone patterns, a similar decision must be taken: should the blocks be laid square to the property, as a 90-degree herringbone or at an angle of 45 degrees? While a 90-degree herringbone is slightly easier to establish, and often results in simpler cuts on a rectangular layout, a

45-degree pattern looks more professional and is the preferred choice of most clients. When setting out a 45-degree herringbone it is best to ensure that the chevrons of the pattern run lengthways (longitudinally) to the property rather than transversely, otherwise the pattern may not be apparent.

## Setting-out a 90-degree Herringbone

Select a long, straight edge to the pavement that will act as the baseline. This may be the front or the side of the house, or some other fairly long, straight line. If none is obvious, create one by establishing a taut string line at some convenient point and extending it as far as possible.

The first block (1) is positioned as a header (lengthways). The second block (2) is rotated through 90 degrees and laid against the first block as a stretcher (widthways). Blocks should be laid 'hand-tight', with the block pressed against the sides of the preceding block(s) and then lowered straight down, ensuring that no laying course material is trapped in the joint. The third block (3) is laid as another stretcher, and then, as can be seen in the accompanying diagram, the next block would need to be a half. Cuts are not fitted at this stage. When block

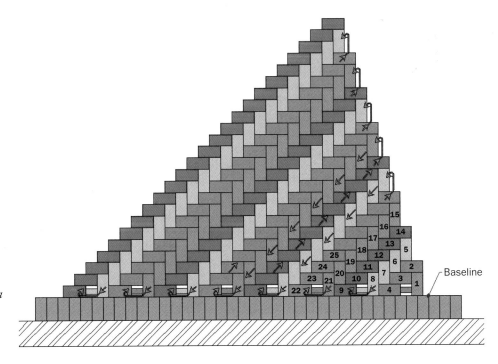

*Establishing a 90-degree herringbone pattern.*

laying, full blocks are laid first, covering the screeded bed as quickly as possible so that it is less likely to be walked upon, disturbed, or visited by the neighbour's cat. The cut blocks are fitted once all of the full blocks are in place. However, when using this pattern, a half-block is necessary to determine the spacing of subsequent blocks. This is achieved by using a brick standing on its end, as shown. This gives the half-brick spacing without interrupting the laying process. Now, block 4 can be positioned to return us to the base line.

Going back 'up' that course, blocks 5, 6, 7 and 8 are laid in sequence, followed by another block-on-end half. Blocks 9–14 are laid as stretchers, moving away from the baseline and then blocks 15–21 are laid as headers returning to the baseline. The laying process continues. Although the pattern being created is square to the property, notice how the laying involves alternating courses of headers then stretchers, working along lines diagonal to the base line and relying on block-on-end halves to check the spacing as work progresses.

## Setting-out a 45-degree Herringbone

This method creates an accurately aligned, 45-degree herringbone pattern, with the chevrons running longitudinally, and positioned so that the blocks that need to be cut against the edge courses will be reasonably large and not the 'darts' or small triangles that should be avoided (see later).

A course of blocks on their sides is laid so that they are tight against the edge course. These blocks are a temporary spacing aid and will be removed once the first few courses are in place. The first block is laid in the normal way (flat), with one corner tight against the blocks-on-edge and aligned at approximately 45 degrees. Next, block 2 is added and aligned to the top of block 1, as shown on page 145. This gives the position for block 3, nestling hard against blocks 1 and 2, which, in turn, determines the position of block 4. The alignment of these first four blocks may need to be adjusted to ensure that the lower corner of blocks 1 and 4 lies against the spacer course, and that blocks 2 and 3 are aligned neatly against them.

This arrangement is repeated with another pair of blocks, numbered 5 and 6, and then blocks 7, 8 and so on, working from left to right, ensuring that the lower corner of the first course blocks lies hard against the spacer course each time, and that the second course blocks are accurately positioned to give square and even joints. This double row, blocks 1–28, creates the first chevron, and note that it is running parallel to our edge course.

Blocks may now be laid in courses running left-to-right and then right-to-left, covering the ground relatively quickly. Note how the third course is started off at the left-hand end, with blocks laid sequentially from left to right, and then the fourth course is laid right-to-left. From this point on, the laying direction of each successive course will alternate, as shown.

Eventually, the blocks laid against the edge course on their sides as spacer blocks can be lifted out and cut to size to piece in, ensuring that the full blocks are firmly held in place.

## Setting out a Stretcher Course and/or a Basketweave

These patterns are essentially 'square'. Stretcher bond, as noted previously, should be transverse to the direction of traffic, and a basketweave is the squarest of the square, being a variation of a chessboard pattern. Both can be set out by using our old friend the 3–4–5 triangle to create a perpendicular to the baseline, and then using that as a guide to the alignment of the first course.

Naturally, where an edge course has been laid parallel to the coursing of a stretcher bond, laying can start against that edge, with no need to establish guide lines. On the project driveway, a narrow pathway linking to the rear garden is being laid with the courses running parallel to the edge courses, as can be seen in the accompanying photograph.

Using standard, rectangular blocks with a stretcher bond, the most visually pleasing effect is achieved with a half-bond, where the joints on successive courses are aligned with the imaginary centre line of blocks on the preceding course. This half-brick offset may be created by using the brick-stood-on-end technique described above. Variations are possible – third-bond or quarter-bond are sometimes used, but they have the effect of leading the eye off at an angle to the main direction of the paving (see the laying patterns illustrated in Chapter 2 on page 19).

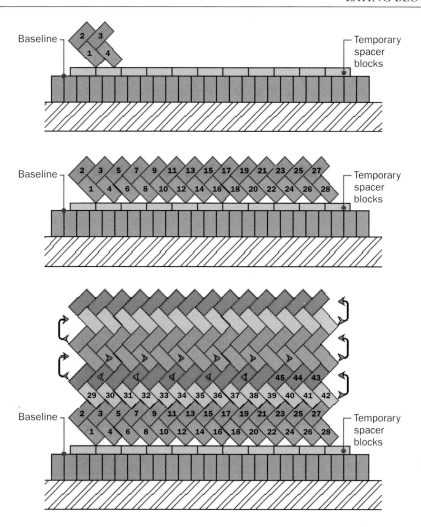

*Establishing a 45-degree herringbone pattern.*

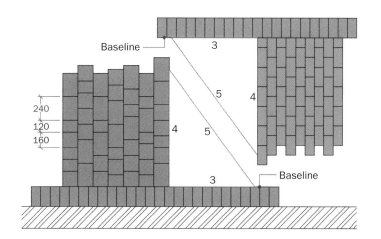

*Establishing 'square' patterns.*

*Laying blocks parallel to an edge course.*

*For the project driveway the first course of blocks is laid to a taut string line projected along the front of the house.*

Many tumbled blocks come in three different sizes, so to rely on a half-bond or quarter-bond is meaningless. The key to laying stretcher bond with these products is to ensure that there is no alignment or concurrence of vertical joints (the perpends) on adjacent courses. It is possible to achieve this by using full blocks, as shown in the diagram, but it is also possible to cheat by offsetting courses by a non-modular distance. For example, one popular tumbled block comes in lengths of 240, 160 and 120mm, so a course offset by, say, 75mm, has no chance of the accidental alignment of perpend joints.

Basketweave is a simple alternating pattern – two headers followed by two stretchers followed by two headers, and so on. Variations are possible by off-setting the subsequent courses by half a block or a full block. Another possibility is to omit one of the header blocks, so that the pattern becomes one header–two stretchers–one header-two stretchers, and so on. This is sometimes known as a California weave.

## The Laying Process

Once the pattern is established laying can proceed at a great rate. A professional laying gang of three will expect to cover at least 100m² each day, and even a pair of DIY enthusiasts should be able to cover 25–30m² per day, although they may well ache as they never have before for a few days afterwards.

As already stated, laying is done from behind the laying edge, that is, the laying operative should be positioned on paving that has already been laid and not on the screeded bed. Consider how blocks will be delivered to the laying edge: is it possible to get them to it without crossing the screeded area?

Unlaid blocks should be kept back from the leading edge by at least half a metre. If they are stacked right at the edge there is a risk of loose blocks at the leading edge being knocked slightly out of alignment. Just a disturbance of a couple of millimetres may be enough to mar a laying pattern and require time-consuming remedial work at a later stage when the error is noticed.

As laying progresses it is important to stand back every few minutes and visually assess the overall layout and alignment. The pattern may drift off course, particularly when the work is more than a couple of metres away from a baseline. It is good practice to establish secondary taut string lines to guide the alignment, to ensure that any drifting is spotted immediately and corrective action taken. In particular, clay pavers, because of their imperfect shape, are notoriously awkward to keep on a true and steady course, so secondary alignment lines are not only recommended, they are essential.

Use a rubber mallet to tap the courses of blocks into line and to keep the joints relatively tight. If joints need to be 'opened up' to maintain alignment, the blade of a spade or a small wrecker/prise bar can provide sufficient leverage to ease out the blocks as far as is necessary. Regular checking of the alignment is vital if the completed pavement is to look anywhere near accurate. Only on very rare occasions does pattern drift correct itself – in actuality, it has a galling tendency to become progressively worse as the laying proceeds, so prompt correction can save a lot of time, effort and tears later on. As a rough guide, an assessment of the alignment every five or six courses usually identifies small molehill problems before they have a chance to become mountains.

Do not waste time paving right up to access covers, linear drainage channels and the like. These can be pieced-in later as part of the cutting-in work. Leave off paving at least one block back from the obstruction and rely on taut string lines to guide the pattern around it in a true and accurate alignment.

## A Few Notes on Laying Techniques

Regardless of the type of block or pattern, it is a common error for laying operatives to take blocks from a single pack of blocks at a time and lay them in more or less the same sequence in which they were packed. This should be avoided since it leads to 'blotching' – sections of pavement having identical colouring rather than a random mix of colours. With single colour blocks, mixing blocks from a minimum of three packs ensures that any minor shade variation is spread throughout the pavement. For multicolour blocks, such as the ever-popular brindle colour, the manufacturing process often results in the colouring being continued from one block to its mould-neighbour, and, if blocks are drawn from a single pack, this colour distribution may be repeated in the pavement, disrupting the random appearance that is required with a multicolour block.

Some blocks, particularly certain brands imported from Europe, achieve a multicolour effect by mixing blocks of three or more specific shades, and, if blocks from a number of packs are not well mixed and randomized before being laid, the result will be very distinctive 'splotches' of individual shades within the completed pavement. The need to mix and randomize tumbled blocks is less critical. As most of these blocks are tumbled in a large, rotating drum, they have already been randomized before being packed. However, there may be some clustering of particular colours, so drawing blocks from a number of packs before their laying will ensure a truly random mix of colours.

Of more concern regarding tumbled and other multi-size blocks, is the ratio between the sizes to be laid. In a three-size mix of blocks, it may seem that a simple 1:1:1 would be the logical choice, but such ratios actually result in what seems to be an excess of the smaller units. Many manufacturers provide 'recommended ratios' for their products and, while these are useful, they should not be regarded as sacrosanct. They are merely suggestions and may not give the appearance you seek. In some cases there is the suspicion that these recommendations are chosen to suit the manufacturers' production, packaging and distribution systems rather than to produce an aesthetically pleasing blend of sizes.

Generally speaking, the largest blocks are preferred. They prevent a bitty look; they cover the ground more quickly and require less jointing sand per square metre, so most laying ratios tend to favour their use. However, when ordering blocks, you should check whether the blocks are delivered in

*Once the first course is established, subsequent ones follow, with the operatives standing on the laid pavers …*

*… and the ground begins to be covered quite rapidly.*

*In just a few hours, almost all the driveway is paved over.*

packs of a single size or packs containing a mix of all the available sizes. For products supplied in mixed-size packs, ordering is simplified and the laying ratios are predetermined, but for those supplied in single-size packs, some consideration is needed. Packs tend to contain the same total area of blocks, but the difference in sizes mean that there will be a different number of blocks in the packs. A pack of 120mm blocks contains twice as many units as a pack of 240mm blocks. So, if a laying ratio of, say, 5:3:1 were to be used, it is not simply a matter of ordering the packs in the same ratio – the number of blocks per pack also needs to be taken into account.

## CUTTING-IN

Once all the full blocks (flags, setts, etc.) have been laid, the alignment should be checked one final time, and then the great fun that is cutting-in can begin.

Essentially, a full block is offered into position, marked where it needs to be cut to fit into the gap available and is then cropped, split or sawn as necessary before being slotted into place. However, as with much else in the paving trade, it is not always as simple as that. There is a fundamental rule that needs to be followed when cutting-in: that no cut piece should be less than one-quarter of a full block. That is the requirement of BS7533, Part 3, but, in practice, pieces smaller than one-third of a block are often deemed to be the minimum acceptable size. So

bearing in mind this requirement, how would the scenario illustrated below be cut-in?

Maintaining the pattern results in a number of small triangles, narrow 'slips', and some extremely small pieces we call 'darts'. These small cut pieces are to be avoided because they are often involved in problems in the medium- to long-term. Thin slips are prone to breaking and one small piece becomes two even smaller pieces; darts sink and force themselves into the bedding, creating small recesses that accumulate detritus and promote the growth of weeds; other small pieces break or come loose and so allow the rest of the paving to move and loosen its interlock.

### In-board Cutting

The solution is known as 'in-board cutting'. This ignores the pattern and uses half-bricks to create gaps requiring pieces larger than the minimum one-third/quarter block. Working out just where these half-blocks are required can be a challenge at first, but, wherever a gap requires a small piece there will usually be some way of removing one of the full blocks abutting the cut and using a half-block in its place.

With coursed patterns, small pieces at edges running parallel to the coursing can be eliminated by turning full blocks (or the largest blocks in a multi-size product) through 90 degrees so that they cover the original course and cross into the subsequent

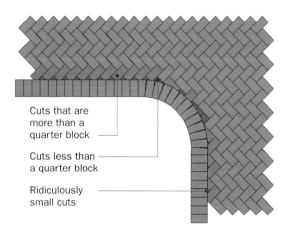

*Example of cutting-in task incorrectly done.*

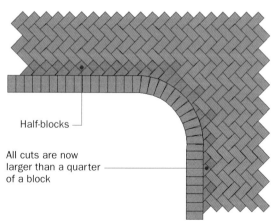

*Cutting-in done correctly with in-board cuts.*

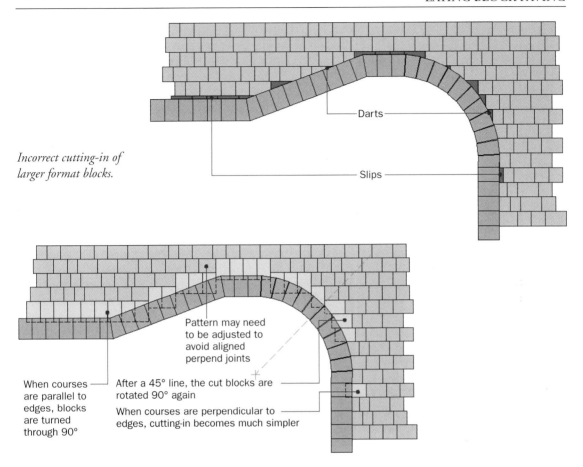

*Incorrect cutting-in of larger format blocks.*

Darts

Slips

Pattern may need
to be adjusted to
avoid aligned
perpend joints

When courses
are parallel to
edges, blocks
are turned
through 90°

After a 45° line, the cut blocks are
rotated 90° again

When courses are perpendicular to
edges, cutting-in becomes much simpler

*Correct cutting-in of larger format blocks.*

course that would have contained the small piece. When cutting in around an arc, once an imaginary 45-degree line has been crossed, the blocks are rotated through 90 degrees once again to ensure that the angle of cut remains less than 45 degrees.

## Marking-up

Marking blocks for cutting can be done in a variety of ways. A small tape measure may be used to measure the gap and transfer the required size to a spare block. A cut-profiling tools might be used or it may be found best to put a block into position and rely on having eyes like an eagle to mark up the line of cut in the correct position. There is no 'right way' – if the chosen method works and results in a cut piece that fits perfectly, then it *is* right.

To mark the line of cut, a pencil, steel nail or the edge of a piece of slate may be used. As long as it produces a clear and accurate mark it will do. How that line is then cut is the next question.

*Blocks are marked ready for cutting.*

149

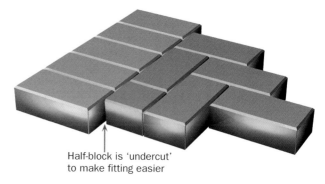

Half-block is 'undercut'
to make fitting easier

*ABOVE: Undercutting makes fitting the blocks much easier.*

*LEFT: The block is positioned between the jaws of the splitter and the lever forced down to snap to block along the desired line (in theory).*

### The Cutting Kit

Traditionally, block splitters were used to cut all blocks, and, while these are fine for straight-line cuts, they can be less reliable for awkward cuts or cuts at relatively acute angles, and consequently many contractors now use a power saw fitted with a diamond blade. While these are accurate and neat, they produce a cloud of dust and are not as quick, in skilled hands, as a splitter. There is also the valid criticism that with tumbled pavers a perfectly neat and crisp cut can look out of place among all those bruised and battered edges, and so some contractors use a saw for standard blocks and a splitter for tumbled pavers.

When cutting, aim to 'undercut' the block slightly. This can be achieved by angling the block when placing it into the jaws of a splitter or by angling the power saw, if that is the tool of choice. Undercutting makes it easier to fit the cut piece into position and also makes it easier to trim back the top face should the piece prove to be a millimetre or two oversize. A cut piece should just drop into the gap. It should not be so tight that it needs to be hammered into place, but neither should it be too loose. When in place, the joint width should not be greater than 5mm and the cut edge should be parallel to the fixed edge. After completing a line of cuts, stand back and assess the line. What might seem to be a slight variation in cut alignment on a single block may look like a dog's back leg when viewed in the context of its neighbours. To paraphrase, 'if a cut block offends thy eye, pluck it out', and cut a new piece.

Training the eye to accurately mark blocks for cutting is something that comes only with practice, and for many DIY projects it is unlikely that there will be sufficient cutting-in work required to accustom the eye fully, so be prepared for frustrations. Some pieces may need to be cut repeatedly before success, but persevere. Bearing in mind that cutting-in can result in a higher than normal wastage rate, it is worth ordering extra blocks to compensate for losses and discards.

## JOINTING AND COMPLETION

Once all the cuts are complete, the joints can be filled and the final consolidation undertaken. For flexible pavements of blocks, flags, and some types of setts, jointing sand is used. As with the bedding sand, this is not just any sand that happens to be available, but is a clay-free, hard, angular sand with grain sizes that generate a high degree of interlock. It is the interlock between sand grains, and between the sand and the sides of the blocks, which imparts a great deal of the load-bearing capacity of the completed pavement.

There is some difference of opinion regarding the best method for jointing. Some sources advocate consolidating the blocks and then sweeping in the sand, while others, myself included, prefer to brush in the sand, sweep off the surplus and then consolidate. It is difficult to say that one method is better than the other, but over many years the 'sand-first' approach has served me well.

## The Kit

Some manufacturers and contractors claim that the forces generated by the steel plate of the vibrating compactor are liable to damage the surface of the blocks, and that a neoprene 'sole' should be fitted to the plate to act as a sort of cushion, minimizing the risk of damage while still transmitting sufficient compaction force to achieve full consolidation. On site, most contractors do not bother with a sole since they have found that the give in the laying course provides sufficient cushioning to minimize any damage.

Some clay pavers are considered to be more brittle than their concrete cousins and more prone to damage, therefore a sole may well be fitted for the consolidation of clay pavers. Alternatively, some contractors strap a piece of old carpet to the plate.

## The Sand

The jointing sand is usually sold as 'Kiln Dried Jointing Sand for Block Paving' and comes supplied in 25 or 40kg bags. It should be bone dry when the bags are opened so that it is free running and will trickle into the joints without delay. Jointing is best done when the surface is dry, otherwise the sand picks up moisture, becomes 'claggy', and no longer runs freely into the joints.

## Brushing-in

The sand is spread over the surface and swept into the open joints with a soft brush, moving to and fro, from side-to-side, until all joints are filled. Surplus sand should be swept off to one side before consolidation; it will be needed again once the plate compactor has done its stuff and so it should not be removed completely nor rebagged, but swept into a pile somewhere out of the way. This is done because some sands are prone to being crushed under the plate and this may cause long-lasting, disfiguring marks on the surface.

## Compacting

The plate compactor is brought on to the pavement and the condition of the underside of the plate is checked, just in case any mud, clay or cementitious material is stuck there. It must be clean before consolidation otherwise the blocks may well be damaged or disfigured. Start consolidation at the edges, with the plate riding half on the edge course and half on the body blocks. Make one pass 'up' the pavement, turn through 180 degrees and return 'down' the pavement, moving the machine over by half a plate width, repeating this at the end of each run and shifting over a half-width each time. When the whole pavement has been compacted, repeat the process transversely so the passes are now being made from side-to-side, again half-lapping with each pass.

The jointing sand may disappear. Joints that moments ago seemed full to the brim now look

*Selected jointing sand is brushed into the empty joints.*

*The plate compactor consolidates the paving blocks and rattles the sand into the joints.*

empty – the vibration of the plate has shaken the sand down into the joints and made room for more material, which will be brushed in once the consolidation work is complete. Consolidation is usually achieved in three or four passes; up, down and side-to-side are usually adequate. Once full consolidation is reached (compaction to refusal), the plate often begins, literally, to bounce around on the surface. This does not always happen since some pavements are better able than others to absorb vibration, but, if this bouncing is noticed, cease consolidation immediately because continuing will damage the blocks.

## Finishing Touches

Once consolidation is complete, the jointing sand that was brushed to one side can be brought back on to the pavement and swept around to refill the joints.

*The finished driveway – compare it with how it looked just a few days before (page 128).*

It is impossible to fill the joints completely and be sure that there will be no further settlement of the sand. Over the coming few weeks it *will* settle and some will be lost due to scour by wind and surface water. Four weeks or so after completion the jointing should be checked and topped up as required. Further inspections may be required at monthly intervals for the first 4 to 6 months, but after that, it is most likely that the joints will be full and stay full. Once the joints are topped up, any surplus sand may be swept off and used elsewhere or rebagged and stored in a dry place until needed. The paving is now ready to be trafficked. No waiting, no keeping off for a few days, no need to give it a rest, you can bring in the car immediately. The paving is complete and ready for use.

## Wet or Damp Conditions

Living, as we do, in these damp islands, there is a reasonable chance that the surface of the paving will not be dry when it is time to do the jointing. During the summer months, when the rain is warmer, a delay of a day or so is all that is needed, but, if you are desperate to complete the paving, brush in as much sand as possible and be prepared to top it up as soon as there is a dry spell.

During the wetter months it may be necessary to water-in the jointing sand; it could be weeks before a dry spell arrives and getting the jointing sand into place is critical to the pavement's performance. Brush in the sand as effectively as possible (when the sand is damp a stiff brush is more effective than a soft one) and use a fine rose attached to a watering can or a hose to wash in additional sand. Try to avoid saturating the paving – the process may be repeated a few days later or loose sand may be left on the surface to allow wind and rain to chase it into the joints, if necessary. Provided that some sand finds its way into the joints, the paving should be satisfactory, certainly for foot traffic and probably for low-speed cars. Avoid heavier cars, 4¥4s and vans, for instance, until the joints are topped up. Tell your partner, children and visitors not to accelerate away at top speed, and to minimize point turns with the power steering until the joints are filled.

# CHAPTER 11

# Laying Techniques for Flags

## INDIVIDUAL BEDDING FOR FLAGS, SETTS AND COBBLES

The previous chapter concentrated on paving laid on to a screeded bed. However, some materials are not compatible with this method, usually because the thickness of the units being laid is too variable to rely on a bed of a fixed depth. Wet-cast patio flags, stone flags, and some setts or cubes are the prime examples of materials not suitable for screed bedding, but, also, there are occasions when regular units, such as block pavers or pressed flags, are laid in small numbers and it is easier to lay them using individual bedding.

### Thick and Thin
As shown in the illustrations, many wet-cast flags for patios and pathways exhibit a significant variation in thickness. Even though the blurb may claim a 38mm thickness, this will be an average and there are likely to be some flags only 32mm thick and some 44mm thick, while other flags may well be 34mm thick at one end and 42mm thick at the other. That is the nature of wet-casting – it is an imprecise technique.

Natural stone flags, particularly the ever-popular

riven flags, are also prone to variation in thickness, as it is often determined by the geology of the parent rock strata. Although they may be advertised as 30mm thick, they could well be anything from 22 to 45mm when they are unpacked from the delivery crates. This does not mean that these items are inferior to pressed concrete products with their more precise dimensional tolerances, but it does mean that laying them on a screeded bed is unlikely to be successful. The only way to accommodate this variation is to prepare each bed individually, tailoring it to the requisite depth and compensating for variation in the base of the unit.

### Common Courses
The preparatory work described previously (excavation, sub-bases, geotextiles and edge courses) remains identical – it is just the laying technique that is different. So, for a project involving the construction of a stone flag driveway, the excavation would be carried out in the way that we saw earlier, as would the construction of any sub-base that was required. Edge courses, if used, could be constructed before the laying or undertaken as part of the laying process. Unlike block paving and other screed-bedded pavings, most patio flags, stone flags and setts or cubes do not rely on restraining edge to hold everything in place.

### Types of Individual Bedding
There are two popular methods for individual bedding: full bed and spot bedding. Full bed involves creating a bed that extends beneath the entire paving unit, supporting each and every part of it. Spot

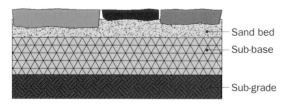

Sand bed
Sub-base
Sub-grade

*The thickness of flags may be quite variable.*

153

bedding involves using four or five dollops of material to support the paving unit only in the corners and possibly the centre too.

Many years of experience in the paving industry have taught that spot bedding is vastly inferior for a number of reasons and is often the cause of medium-to-long term problems with the paving. It is often illustrated in basic 'how to' guides, and in some manufacturers' catalogues, since it is believed that spot bedding is simpler and more forgiving for the inexperienced flag layer. It is in the manufacturers' and the retailers' interests to promote the notion that flag laying is incredibly simple, and there is no doubt that spot bedding is inexplicably popular among the inexperienced. However, because the flag is supported only on a number of fixed points, it requires just one of those points to settle or move very slightly and the whole flag becomes unsteady and starts to rock.

If spot bedding using a mortar bedding material is done in such a way that, when the flag is tapped down to level, the spots or dollops of mortar spread and coalesce into one large mass, there is less of a problem, but in many cases the spots are positioned and sized in a manner that precludes such merging and the flag ends up being supported at just four or five points. So it should come as no surprise that the remainder of this chapter concentrates exclusively on the full bed method. This is the method required by the British Standard covering flag laying (BS7533,

Part 4) and is the only method ever used by my paving company.

## Choice of Laying Course Material

A full bed can be prepared by using clean (that is, cement-free) sand, a mortar or a concrete. Different laying course materials are selected for different projects. Clean sand may be selected for a temporary patio or a pathway, for pavements that are expected to move slightly over time (such as many public footpaths), or for some vehicle-overrun pavements. Mortar bedding is most popular for patios where a fixed and firm but low-strength laying course material is required, and concrete bedding is most commonly used for driveways and other pavements that will be required to carry vehicles but are not suitable for an unbound laying course material. The general advice is that flags smaller than $450 \times 450$mm are laid on sand; those larger are laid on mortar or concrete.

Where sand is used, it should be a grit sand as used for block paving. This is used for exactly the same reasons as it is beneath block paving, etc.: it is free-draining and less likely to be washed out by ground water. Soft sands are too prone to being washed out, are at risk of becoming 'mushy' when saturated, are water-retaining rather than free-draining, and can provide a home for vegetation and weeds.

Mortar bedding is usually a relatively weak material. The cement content is included to resist wash-out and to bind the sand content so that it permanently supports the paving units. For flagstones, a mix ratio of 10:1 is adequate since the laying course is not required to be incredibly strong – all it needs to do is provide adequate support for the paving. Setts, cubes and other, smaller paving units may require a richer mix, something like 6:1. A grit sand should be used to prepare the mortar because this helps to ensure that the laying course will be free-draining and permeable, even with the cement content. The use of soft or building sand can result in an impermeable laying course, which may not always be desirable.

Concrete bedding is also a relatively weak mix, an ST1 concrete (1:3:6) is usually adequate for most driveways. Concrete is selected in preference to mortar bedding when the depth of the laying course

---

### Why Spot Bedding Should Be Avoided

- Voids are left beneath the flags, making them more liable to break;
- Voids allow water to collect, which can cause settlement;
- Voids provide living space for small mammals, ants and other mining creatures;
- One spot settles and the whole flag starts to rock;
- To fix any rocking flags requires the mortar spots to be broken out;
- It is often more expensive than a full bed;
- It does not comply with the Code of Practice for laying flags.

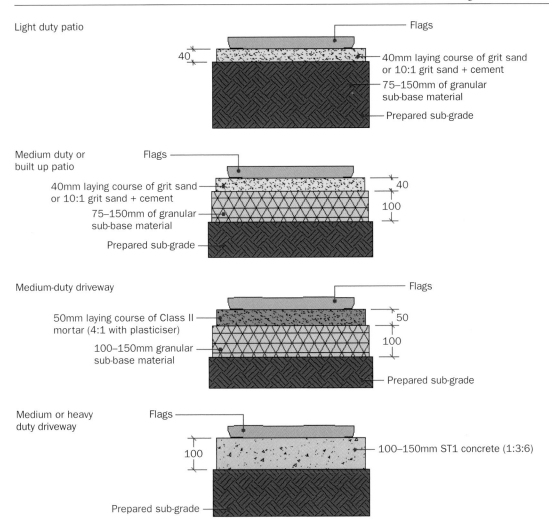

Light duty patio

Flags

40

40mm laying course of grit sand or 10:1 grit sand + cement

75–150mm of granular sub-base material

Prepared sub-grade

Medium duty or built up patio

Flags

40mm laying course of grit sand or 10:1 grit sand + cement

75–150mm of granular sub-base material

Prepared sub-grade

40

100

Medium-duty driveway

Flags

50mm laying course of Class II mortar (4:1 with plasticiser)

100–150mm granular sub-base material

50

100

Prepared sub-grade

Medium or heavy duty driveway

Flags

100

100–150mm ST1 concrete (1:3:6)

Prepared sub-grade

*Various bedding and laying options for different types of flagged pavement.*

is likely to exceed 50mm or when regular vehicular traffic is expected.

For both mortar and concrete bedding, the slump or 'wetness' of the mix is usually a matter of personal preference. A semi-dry mix is often the easiest to work with, but requires the greatest degree of accuracy when it comes to the bed preparation since the material cannot 'flow' as does a wet mix. Wet mixes flow, but they also have a tendency to cause the flags or other paving units to 'float', and, when a neighbouring unit is being tapped down to level, a unit already laid may be 'pumped up' by the

movement or flow of the wet laying course material. A moist mix is a happy medium.

It is worth pointing out that there is no need for the flags to bond or to stick to the bedding, except for when they are laid at a free edge, or laid as steps and copings. For flags laid as a simple pavement, there is nothing to be gained in having the flags stuck to the bedding, and, if it ever became necessary to lift the flags, any attached bedding would hinder their extraction and would probably mean that someone would have to spend hours chiselling off the mortar or concrete before they could be relaid.

# THE LAYING PROCESS

Unlike screed bedding, individual bedding does not have its level predetermined by edge courses, screed rails and a screed board. Instead, the level of the bed is established provisionally and it is the surface level of the paving that is used to determine accuracy. Accordingly, some method of checking the level at any point on the pavement is required. This is usually done with a series of taut string lines, stretched from one known, level reference point to another, and then a straight-edged timber and spirit level will be used to cross-reference the string lines. Taut string lines are also used to guide alignment at free edges. Once the string line level guides are in place, laying course material is spread out, roughly levelled and lightly compacted, either by trampling it underfoot or by one or two passes with a plate compactor. At this stage the laying course material does not need to be fully compacted since some give is needed to accommodate the eventual paving.

## Making the Bed

The paving unit to be laid is examined and its thickness assessed. A bed is prepared for it, using a spade, trowel or float to level out the bedding material, aiming to mirror any variation in thickness. The bed is 'rippled' so that there will some give when the unit is laid and tapped down to level. Judging the level of a bed is part art and part science and something that becomes easier with practice. Good flaggers and sett layers can judge the thickness and requirements of a flag or a sett in a single glance and prepare the perfect bed in a matter of seconds, but it may have taken them years to learn this skill. For most people, bed preparation is largely a matter of trial and error.

## Buttering

When laying flags that will have mortar joints, the 'receiving edges', that is, those edges on existing flags, kerbs, edgings or walls that will abut the flag when it is eventually laid, are 'buttered' with a general purpose mortar. The mortar is applied to the receiving edges in generous quantities, but without getting any on to the surface. As the flag is laid, it presses against this mortar, forming a well-filled joint between the two units that will eventually be topped up with pointing mortar. Buttering also has the benefit of regularizing the width of the joints: it will not make every joint exactly the same, but it will make most of them very similar in width.

## Laying

Bed and butter are ready, so the next task is to lay the flag. Smaller flags (anything under 450 ¥ 450mm) can usually be lifted into place direct and pressed into both the bed and the butter mortar in one operation. Larger flags need to be lowered into place, either from the existing paving or from the bedding area. Unlike screed bedding, there is no requirement to avoid stepping on to the laying course material before it is covered with the paving: it is perfectly acceptable to lay from this leading edge.

## Consolidating

The flag is lying on top of the prepared bed and is ready to be tapped down to level. This is best done with a pavior's maul, although smaller flags laid on mortar or concrete can often be consolidated by using a rubber mallet. It is important that the consolidation is done at the correct places on the flag's surface. Tapping down on the corners is more likely to flick the flag out of its bed, and simply hammering it in the centre is unlikely to result in the flag being consolidated evenly across the entire bed. The 'targets' for the tapping down are four imaginary spots located on the diagonals of the flag and halfway between the centre and the corners and the imaginary rectangle connecting these four points. Work around this imaginary rectangle, moving from point to point, tapping down until the flag will go down no

---

**A Recipe for Buttering and Pointing Mortar**

A general purpose mortar for buttering and pointing comprises:

- 4 parts of soft or building sand
- 1 part of cement
- dye or colour, as required
- plasticizer or small squirt of washing-up liquid
- clean water

---

*An individual bed is prepared by levelling out the laying course material with a shovel or a trowel.*

*Mortar is 'buttered' on to the receiving edges of already laid flags.*

*Small flags can be lifted into position.*

*Larger flags should be carefully lowered into place.*

*The rubber mallet is used to consolidate the flags.*

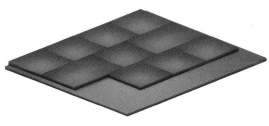

*Consolidation points – aim for these!*

further and is flush or level with its neighbours. The level may also be checked against a string line guide and/or by using the straight-edge and spirit level.

### Too High

Sometimes the flag will not consolidate any further, but is still high, relative to its neighbours and the level guides. Repeated hammering is likely only to break the flag. If it won't go down, then it won't go down – the only remedy is to lift it out, set it aside and create a new bed. Make a mental note of just where it was high, so that the existing bed can be reduced in that area. Loosen the top 12–15mm of the bed, replace the butter mortar if necessary, try the flag again and hope that it will go down this time!

### Too Low

At other times the flag keeps going down, below the required level. The bed is too low and this can be rectified only by lifting out the flag and adding more laying course material to the existing bed. Again, make a mental note of areas needing attention and adjust the bed accordingly by sprinkling on additional bedding. Loosen the top 12–15mm of the entire area with the point of a trowel, freshen up the butter mortar and re-lay the flag. Lifting and re-laying flags is hard, tiring, back-wrenching work and reduces the amount of paving completed in any session, so getting the bed right first time is the target. Not even the best flagger in the world gets it right first time every time, so do not get too

---

## Which Way Up?

There is some confusion as to which way is up with some flags, especially the imported sandstones and limestone flags from India. There are two key indicators to which is the top surface, and both relate to the way the edges are formed. Stone flags are cut with sides that taper inwards, so that the upper face, the one presented to the world and walked upon, is slightly larger than the base. Look at the sides of a flag, edge on, and notice that it is not square, but slopes inwards. This is done to ensure that an adequate portion of mortar is trapped between adjacent flags and to ensure a relatively tight joint at the surface. Secondly, the upper face will normally have dressed arrises – the edges or vertices between the sides and the upper faces. Again, this is done to present a neat and tidy joint when the flags are laid. Conversely, the base is often haphazard, jagged, irregular and untidy.

*Edge-on view of stone flags. Note how the edges are tapered – the top face is always larger than the base.*

LEFT: *The upper face of a stone flag has neat edges and an attractive face …*

RIGHT: *… while the underside has more ragged edges and an obviously uneven surface.*

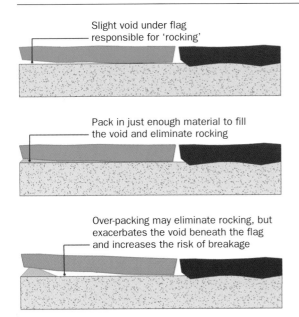

*Be careful when 'packing' loose or low flags.*

frustrated if the bedding is not working for you. As more and more flags are laid, the ability to correctly gauge the required bed will improve, and you will find that remedial work to the bed is required less and less often.

*Checking*

When the flag is in place and at the correct level it needs to be checked to ensure that it is not loose nor rocking. Some smaller units will appear to rock, even if firmly bedded, by reason of their small size, so the rocking test may not be necessary on smaller flags. But for larger units the rocking test is essential. Straddle the flag along one of its diagonals and then shift your weight back and forth to see whether the flag moves, then change and test the other diagonal; there should be no obvious movement in the flag.

Any slight movement can be eliminated by 'packing' one of the moving corners, but care is needed to ensure that any such packing does not result in the flag being jacked up off its bed and left hollow beneath. Any rocking movement that cannot be eliminated by a minimal amount of packing is best resolved by lifting the flag, creating a new bed and re-laying it.

## Regularising the Joints

As with all other forms of paving, it is important to stand back from an area that has just been laid, and assess the layout. The surface should be even, with no lips between adjacent units, no high spots and no hollows. It is often noticed that joints need to be 'evened out' to ensure regular joint widths. This is particularly true with modular, random layout flags that are supposed to work together to simplify patterning and minimize cutting. The problem arises because, for example, two 300 × 300mm flags do not add up to one 600 × 300mm unit – there has to be a joint, and if that joint is of the standard 12mm width, two 300mm units actually tally up to 612mm, and three of the 300mm units do not equate to a convenient 900mm but to 924mm.

Apply this knowledge to a larger area of paving, and it becomes apparent that joints will have to vary in width to even-up the alignment, otherwise some joints will be 25mm or more in width, and suddenly there are eye-catching strips of mortar between the flags instead of discreet and tidy pointing. The use of buttered edges during the laying process helps to regularize joints to around 12–15mm, but once a larger area has been covered, individual flags will need to be shunted one way or the other to create even joint widths. This may mean tightening up 12mm wide joints to, say, 8mm, or possibly widening them to 16mm. The tolerance range is

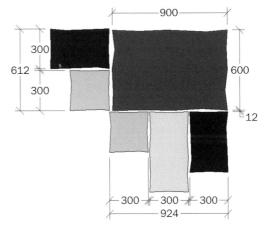

*Joint widths need to be monitored to keep spacing regular.*

159

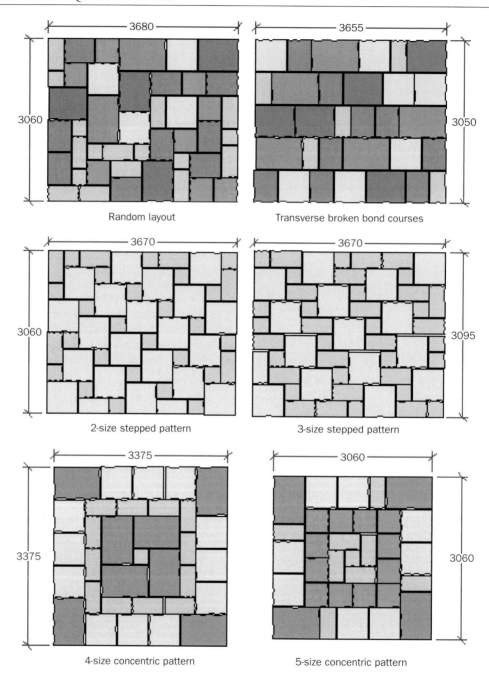

Random layout

Transverse broken bond courses

2-size stepped pattern

3-size stepped pattern

4-size concentric pattern

5-size concentric pattern

*Possible flag laying patterns for modular flags. Random layouts should never have four corners meeting and the straight-line joints should not run for more than 3m. Broken bond coursework has staggered perpendicular joints. There are thousands of potential patterns and layouts: the only limit is the number of flag sizes available in any particular range and your imagination.*

reckoned to be 10–15mm, but, in reality, this is often stretched to 8–18mm.

For flags that are not mortar pointed, joint widths can be regularized during the laying process by inserting small pieces of 3mm (⅛in) tile or board between flags. Some flags are designed to be laid 'butt-jointed' and may have spacers incorporated into their edges to create regular 3mm joint widths. The relative tightness of the joints with these flags is often critical and to open out the joints may adversely affect the structural integrity. With such flags, any required adjustment should be achieved by tailoring the size of cut pieces at the ends of each course, rather than by opening out joints.

## Cutting Flags

In general, the guidance given previously for cutting block paving applies equally to cutting flags. Most cutting is done with a diamond-bladed power saw, but, traditionally, a lump hammer and a pitching chisel would be used to cut flags, and the ability to work both stone and concrete flags with these basic tools was a large part of the mason pavior's art, a skill that is now largely lost.

Cutting with a power saw is straightforward: the unit is marked and the saw is applied. Do not force the saw: allow it to eat into the flag under its own weight while you guide the alignment and the positioning of the blade on the flag. The more the blade is forced into the flag, the less effective it is likely to be. Inboard cuts are used as they would be for block paving, with the aim of eliminating slips, darts and any pieces less than a quarter of a flag.

Where flags need to be reduced in size to fit a particular layout, they should be exchanged for smaller units wherever possible; so, for example, a patio laid with 600 × 600mm flags has one course with a 250mm gap at one end. Rather than slice a 600 × 600mm unit into a 350mm and a 250mm piece, it is better to cut off 50mm from a 600 × 300mm unit.

Other cuts commonly encountered with flags are described below.

### Curves

Curved cuts may seem impossible to create with a power saw since it is difficult to 'steer' the blade around relatively sharp arcs. Happily, there are two solutions. One is to use a curve-cutting blade, which is divided into a dozen or so sections, each aligned at an angle to the previous section. It allows the blade to follow an arc much more easily. The second method is a little more crude but works well for most purposes – it relies on making a series of cuts at different angles, rather than one single cut. (See the diagram below.)

### Notches

Notched cuts are generally used to fit around other surface units, such as stop tap boxes and hoppers. However, any internal cut requiring the removal of more than about 25 per cent of the flag seriously weakens it and makes it likely to break. This can be pre-empted by cutting the flag along a 45-degrees-line from the internal point of intersection; the method recommended by BS7533, Part 4. However, the angled cut line is not always aesthetically pleasing

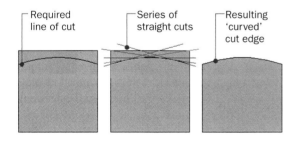

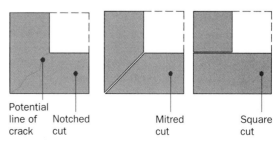

*Cutting curves and notches in flags.*

*Accurate cutting of curves can be essential. (Acheson-Glover)*

and so it is sometimes better to extend one of the cut lines to produce two rectangular pieces of flagstone. (See the diagram on page 161.)

*Diagonals*
Diagonal fold cuts are used within flagged areas that have a significant change of gradient. Due to the larger plan size of flags, laying them as complete units would result in one corner being proud or sticking up and presenting a trip hazard. By slicing the flag along a diagonal, the proud corner can be tapped down to the required level and any hazard eliminated.

## POINTING AND JOINTING

For some reason, this final task in the flag-laying procedure is the one most dreaded by many DIY-ers and contractors alike. A colleague describes it as 'mortarphobia' – the fear of making a mess of all those lovely flags you have just spent two weekends laying. Pointing makes or breaks a flagging project. Done properly and neatly, no one ever notices it, but

make a mess and it is the only thing anyone ever sees, no matter how much was paid for the flags.

There are several different ways in which the joints may be filled. Mortar pointing is one that springs to mind, but there is also 'sand-jointing', 'dry-grouting', 'wet-grouting' and the option to use specialist (pronounced: 'expensive') products that give neat and tidy joints with no risk of staining. Sand-jointing needs no explanation since it is identical to the procedure used with block paving. Brush dry jointing sand into gap, retreat to a local hostelry, and remember to top it up a few weeks later.

### Mortar Pointing
Traditional mortar pointing is the most popular technique. Mortar is packed into the joint and smoothed off. Most problems arise when too wet a mortar is used. The tidiest method is to point the joints immediately after laying, while the butter mortar is still reasonably fresh and unset, using a very, very stiff pointing mortar that has just enough added moisture to bind it together, and no more. The pointing mortar is pushed into the joint from the blade of a larger brick trowel, using either a small pointing trowel or a pointing bar tool, and 'tooled' (polished) to give a smooth, neat finish. As the butter mortar is still fresh, a chemical bond is formed between it and the newly added pointing mortar and the whole will harden as one.

This technique may also be used when there is a delay between the flags being laid and the eventual jointing. As the original butter mortar will have hardened, any bond between old and new mortar will not be as strong, but, provided that there is sufficient depth of fresh mortar, it should be adequate, even for driveways. A slightly wetter mortar may be used, but extra care is needed to ensure that the edges of the flags are not stained when the wetter mortar is pushed into the joints.

All mortar pointing should be done when the flags are dry and no rain is forecast for at least 24hr. Rain or damp surfaces are the enemy of neat and tidy pointing. However, searing summer heat should also be avoided since it dries out the mortar before the water in it is able to hydrate the cement, resulting in a seriously weakened mortar that may well crumble to dust. On hot days leave the work until late afternoon (4pm or later), or cover the freshly pointed joints to protect them from direct sunlight. If it is a hot day and the joints are dry or partly filled with hardened buttering mortar, put a hose to the flags an hour or so beforehand, or use a watering can to saturate the joints. This will 'quench' the flags and the old mortar and so prevent them from 'parching' the fresh mortar when it is applied, by drawing out too much of the vital water content.

The secret of neat and tidy mortar pointing is patience; it is a slow, tedious job and the more it is rushed, the more of a mess it is likely to be. Accept that it may take as long to point the flags as it did to lay them. Professional contractors may well be able to point a 30m² patio in a couple of hours, but that is because they have had years of practice. A typical DIY-er should not be surprised to find that it takes them a full day or even a couple of days to point the same area.

## Dry Grouting

This technique is popular because it is seen as a stain- and risk-free process. However, it can result in a substandard joint unless it is done properly. The basic premise is simple enough: exclude the water from a mortar and there is no risk of staining; mix sand with cement but do not add water, and use a soft brush to sweep the dry powder into the joints. Brilliant!

The problem is that cement needs water to hydrate, to promote the chemical reaction that makes it go hard, and if no moisture is available or the cement has to rely on absorbing a few drops of dew and atmospheric moisture, the hydration process is dragged out over an extended period, resulting in a pathetically weak bond. To counter this, some people brush in the dry sand and cement mix and then use a fine rose on a watering can or on a hosepipe to sprinkle the area with water. That solves the problem regarding lack of water… however, it also has the unfortunate side effect of splashing some of the powder out of the joint and on to the surface of the flag. Some splashed mortar powder may be washed off by more water, but on most jobs some gets wetted on to the flags and, when everything dries out, the flags are left with a white haze or 'frosting' of cement. The severity of this haze varies: on some jobs it may be insignificant and will be weathered away in a

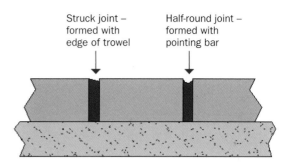

*Tooling joints makes them more resistant to weathering.*

*Use a pointing to feed mortar from a larger brick trowel into the joint …*

*… then use the pointing bar to smooth the mortar.*

*Dry grouting relies on brushing-in a dry mix of sand and cement.*

*Wet-grouting uses a soup-like mortar slurry to fill the joints.*

matter of weeks, but on others it can be a serious problem. The only advice one can give to minimize the effect is to take exceptional care with the application of the water, aiming to water the flags and allow it to trickle *into* the joints, rather than sprinkle it direct *on* to the powder-filled joints.

Dry grouting does work well when it is used to fill joints in which the buttering mortar is less than a couple of hours old and still has a significant moisture content. The dry powder draws water from the existing mortar, the cement hydrates properly and a hard mortar is created. The key to success with this method is to brush in the powder as soon as possible after laying. Furthermore, the dry mix must be firmly pressed into the joint: brushing in and allowing the powder to settle under its own weight will result in a weak, airy, short-lived mortar. The powder needs to be pressed down with the edge of a trowel or a pointing bar, topped up and pressed down again and again until the joint is filled to capacity. The bar or the trowel should be used to strike or polish the joint since this draws moisture towards the surface, helping to activate the essential hydration process.

## Wet Grouting

Bearing in mind the problems with dry grouting, the technique known as wet, or slurry, grouting goes to the other extreme and relies on adding far more water than would normally be used for a mortar. In effect, the mortar is made to the consistency of a hearty pea soup, so that it flows and slops about. This soupy preparation is swept over the entire surface, allowed to flow into the joints and then the excess is swept away.

There is no doubting that this technique normally results in well-filled joints and a good quality mortar. However, it also tends to leave a cementitious haze over the surface. On a hot day, the thin film of mortar soup remaining on the surface dries so quickly that the cement has little chance to hydrate and it can be brushed off after a few minutes. This tends to leave a dusty-looking surface, but, since this technique is mostly used on commercial paving projects where pedestrians and/or vehicles will soon wear away any dusty deposit, it is not considered a major problem. On damp or cooler days, the surplus may be washed off with clean water, although this sometimes results in washing through the cement content, leaving the top 6 to 10mm of the joint as a sandy, weaker mix.

Wet grouting is occasionally used on a driveway or a patio. It works reasonably well with pressed concrete flags and with the smoother stone flags, and it is the most popular method for jointing setts, cubes and cobbles, where the quantity of jointing per square metre is considerable. It is something of a gamble for the slightly more porous, wet-cast concrete flags and for imported sandstone flags with their freshly riven, unpolished surfaces since it can leave a disfiguring cement haze that may take years to disappear fully. It is extremely fast and easy, which is why it is popular for commercial paving projects, but it should be used with caution on patios, paths and driveways unless the paving materials are suitable.

## Alternative Methods

Knowing full well that pointing and grouting are major headaches for designers, contractors and DIY-ers alike, many approaches have been developed over the years in an attempt to simplify the task and give a cleaner finish. Some are more successful than others, but all share a common theme: they are more expensive than one of the methods outlined previously.

### Gun-applied Mortars

The essential idea is that mortar can be 'squirted' into the joint by using something akin to a mastic gun. The mortar is fed to a nozzle, directed into the joint and none need spill over. Some gadgets use a mastic tube cartridge filled with a mortar prepared by using a magic ingredient that is alleged to maintain 'flowability' as it is pumped into the joint. Other systems use a mechanical pump to transfer the magic mortar from a mixing pan along a hose to the gun nozzle. The efficacy of these systems varies and it is not possible to review everything on the market. Some work well, some work, and some are a waste of time. None are particularly cheap, but, factor in the cost of replacing mortar-stained paving and the manufacturers can make a more-or-less plausible economic case. The mortars produced are often of the highest quality, with superb compressive, flexural and tensile strength, and a high bonding factor, but are these qualities really needed on a residential patio or driveway?

### Polymeric Jointing Compounds

Imagine a dry, sandy powder that could be brushed into joints where it would harden of its own volition

*Polymeric jointing compounds can create firmly filled joints with no risk of staining.*

and with no risk of staining. Is this a fantasy? No, for there are now several such products on the market. They rely on a special ingredient that is mixed with a sand or other fine-grained aggregate. When exposed to the atmosphere, the compound hardens as if by magic. They really are as simple as that – brush them in and sit back while they 'set'. The result is a hard material that has filled the joint and left not a mark on the paving. Their drawback is cost. One popular product costs around £20 for a 10kg pack that is sufficient to fill the joints within approximately $4m^2$ of typical, wet-cast, riven patio flags; this cost of about £5 per $m^2$ is to be compared with the equivalent quantity of sand and cement mortar at around 35p per $m^2$.

Furthermore, some products do not actually bond to the paving. The special polymer mixed with the sand is capable of binding only to itself, so that the product forms a 'plug' within the joint, rather than a properly sealed joint. This makes such products unsuitable for those pavings that rely on the jointing for much of their strength, notably setts, cubes and cobbles. If you are considering using a polymeric jointing compound, find out whether it bonds to the paving or only to itself. The latter cannot be recommended for any vehicular areas.

165

# CHAPTER 12

# Laying Other Materials

Most other paving materials are laid by following one of the methods already described, using either a screed or individual bedding. There may be slight variations to the detail, but the basic processes remain the same. Some flags, particularly small-element, pressed concrete flags, are laid in the same way as flexible block paving, on a screeded bed, with sand joints and consolidated by using a vibrating plate compactor. Rigid block paving can be laid on a screeded mortar bed with buttered and pointed joints. Setts and cubes are suited to either method, while cobbles are usually pressed into a high-strength mortar screed. The laying techniques used with these natural materials are considered next.

## SETTS AND CUBES

The characteristics of the setts and cubes to be laid will determine which laying method is most suitable. Units of regular depth could be laid on to a screed, but if there is significant variation in depth among them, individual bedding is more likely to be the preferred method. Cubes, by definition, tend to be of

regular depth. Whether they are new or reclaimed, it is likely that they will have a depth tolerance of around ±10mm. New or pre-sorted setts may be of regular depth, but many reclaimed setts are a jumble of depths and individual bedding is the only suitable laying method.

## Flexible Construction

Some cubes are laid flexibly on a sand or grit bed, the joints filled with grit or a crushed stone dust known as 'splitt'. The laying process is identical to that used for block paving, although more attention needs to be paid to alignment as there will be some variation in plan dimensions. Coursework needs to be laid to taut string guide lines, and the joint width needs to be kept tight. Because of natural variation in the width of setts and cubes, only one 'edge' can be aligned, which should be the front or leading edge.

Arcs and bogens are formed by using cubes and, although complex to set out, they are visually stunning. The joint width varies and may be quite open in places, so the use of a suitable jointing medium becomes essential. The radius of arcs and

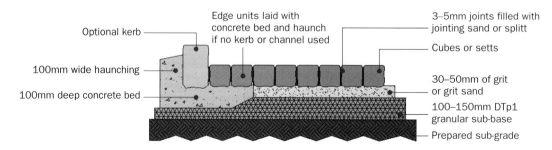

*Cross-section for flexible construction of sett paving.*

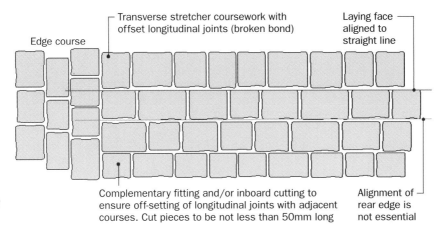

Edge course

Transverse stretcher coursework with offset longitudinal joints (broken bond)

Laying face aligned to straight line

*Align only one edge of the setts in each course.*

Complementary fitting and/or inboard cutting to ensure off-setting of longitudinal joints with adjacent courses. Cut pieces to be not less than 50mm long

Alignment of rear edge is not essential

bogens can be chosen to suit the job in hand, to ensure a full number of arcs span the pavement and that they are evenly spaced.

For flexibly laid setts and cubes, restraining edge courses are essential. These are usually in the form of a channel of setts or cubes laid longitudinally on a bed of concrete, although kerbs, edgings and channel stones are also used. Once an area of paving has been

Fan detail

R1125mm

8–12mm joint width to be maintained as far as possible

Setts need to be cut to fit

Typical layout

4500mm

*Although they look complicated, to set out arcs or bogens is quite easy, if you have a plan.*

laid, the splitt is swept in and the plate compactor used to consolidate the lot. This method of construction is popular in continental Europe.

## Rigid Construction

*Using a Screeded Bed*

For some reason, many sett and cube installations in Britain and Ireland use a rigid bed and rigid jointing. Much the same method as outlined above is used, although the sand bed is replaced with one of mortar or concrete. Once the setts are laid, a clean, 6mm gravel is scattered over the surface to partly fill the joints and hold fast the units while the plate compactor does its job. On completion of an area, a 'slurry grout' (see page 168) is applied over the surface to complete the jointing.

*Fan-pattern setts being laid to a private driveway. (Original Stone)*

Jointing option 1:
Joint filled to within 25mm of top with 6–10mm clean gravel and then filled with hot-poured sett jointing bitumen

Jointing option 2:
Grit sand mortar (3:1 mix) made as a slurry and swept into joints

Jointing option 3:
Proprietary pourable cement-based jointing compound, used to fill joints in accordance with manufacturer's instructions

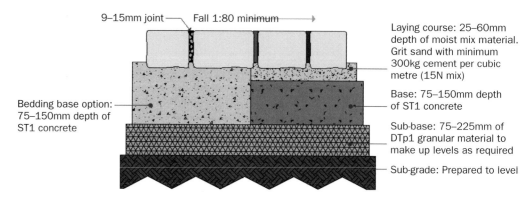

9–15mm joint — Fall 1:80 minimum ⟶

Laying course: 25–60mm depth of moist mix material. Grit sand with minimum 300kg cement per cubic metre (15N mix)

Base: 75–150mm depth of ST1 concrete

Bedding base option: 75–150mm depth of ST1 concrete

Sub-base: 75–225mm of DTp1 granular material to make up levels as required

Sub-grade: Prepared to level

*Cross-section for rigid construction of sett paving.*

### Using Individual Bedding

Where setts of varying depths are used, individual bedding is the only viable laying option. A reasonably strong bedding mix is prepared, 6:1 semi-dry or moist mortar, or an equivalent fine concrete. When laying setts and cubes in this manner it is necessary to work from the unpaved area, so sufficient bedding for just one or two courses will be spread out at any one time, and trampled lightly to partly compact it.

The setts are sorted into groups of similar widths, so that the courses remain parallel, and are brought to the laying face in barrows. The mason selects a sett, assesses its depth and prepares a suitable bed by adding or removing bedding material with a trowel.

*Special jointing mortars can be poured into the joints from a watering can.*

The sett is offered into position and tapped down to level by using a rubber mallet or a pavior's maul. Unlike block paving, courses are completed by finding a sett to close the course or cutting one to suit, before the next course is laid.

## Slurry Jointing

This is probably the most popular method for jointing setts, particularly for reclaimed or new setts laid on a rigid bed. It is essentially the same as the wet-grouting technique described in the previous chapter for use with flags. The slurry mix is brushed into the joints and then the excess is swept from the surface. The same pros and cons apply – the joints so created tend to be well-filled and of good quality, but the surface is often stained or 'hazed'.

Contractors use a range of techniques to minimize the effect of hazing. The slurry is repeatedly swept until the surface is dry and dust-free; sawdust is sprinkled over the slurry residue and then swept off; a cement-retarding agent is sprayed over the surface so that the whole can be washed down with a hose some hours later, cleaning machinery is used to 'wash' the surface one section at a time; and the finished surface may even be sand-blasted to clean it.

For small residential projects, washing off with water is probably the easiest option, although some success is possible by using the watering-can technique, as long as the slurry is continually agitated to prevent settling of the heavier particles while the slurry is in the can.

## Using Pitch

Up until the Second World War, most sett paving in Britain and Ireland was jointed with pitch, a form of tar or bitumen, and there are still thousands of streets performing perfectly well with pitch joints that were poured over a century ago. Then we changed to cement-based jointing, and many post-war installations were plagued with problems, most of which can be attributed to the use of understrength, rigid, cement-based jointing and bedding.

Pitch has matured somewhat over the years, and is now known as 'modified bitumen', meaning that it has been tweaked and fillers added to create a product less prone to melting on hotter days while still providing support from a flexible medium that retains its ability to reseal itself following any minor movement of the setts. The modified bitumen is melted in a tar boiler and poured into the joints from a steel can. It sets as it cools, and any spillages or splashes are easily peeled clean from the surface. There are some problems regarding the use of pitch on sett pavements carrying heavy road vehicles, but for driveways and other residential projects, it is hard to come up with a better looking jointing medium that should give decades of faultless service.

## Traditional Pointing

This one is easily summarized in one word: don't! A typical square metre of 100 × 100mm cubes contains around 16 linear metres of joints. Your knees and your back will never forgive you if you attempt this task by hand, but, if you are a glutton for punishment, make sure that you are using kneepads, allow plenty of time to complete the work and jot down the number of a good chiropractor …

## LAYING COBBLES

Nowadays there is not much call for cobbles, or duckstones as they are called in some parts. They are difficult to walk across, hence their use as 'deterrent paving' in areas that need to be kept free of

*Molten pitch is poured from a spouted steel bucket and requires a great deal of care.*

pedestrians, and they give a harsh rattling to any car driven across them at more than walking pace. However, they are still used in gardens as decorative or feature areas, and there is some demand for them to be relaid within areas that have deteriorated or broken up over time. They are also used on heritage projects, although modern laying methods bear little resemblance to the way they might have been laid originally.

Cobbles were always a low-grade paving. They are slightly preferable to walking through mud, but they were never as grand as setts or flagstones, which are flatter, smoother, and more easily trafficked. Accordingly, traditional construction methods were quite crude, with many cobbled surfaces being created by hammering the stones into hard clay or a lime and clay bedding layer. Modern installation sets them into a mortar or concrete.

For areas that might be required to carry vehicles, a concrete base is required, but pedestrian pavements normally rely on a 'generous' 60 to 100mm-thick bedding layer. A wet mix is normally used, since it should 'flow' around the cobbles when they are inserted, to form an effective seal that will hold them firm and prevent their being loosened. A Class II general-purpose mortar might be used or a similar concrete, something like a 1:2:4 mix. The bedding material should be spread over an area not wider than the reach of the laying operative and then tamped and floated to a flat and reasonably smooth finish. The cobbles are pushed into the mortar or concrete

*The biggest use for laid cobbles is as decorative features.*

to around half their depth and jiggled or lightly tapped to encourage the disturbed mortar or concrete to flow back around them.

Cobbles should be placed as close as possible to their neighbours. The mortar/concrete bed is not particularly attractive, so selecting the correct sizes to create as dense a coverage as is possible becomes the challenge. Once an area of around 600 × 600mm has been covered, a plywood board is placed over the cobbles and tapped gently with a mallet or a hammer, with the intention of levelling-up all the cobbles beneath and encouraging the mortar/concrete to find a uniform level while also sealing-in the stones.

Where the bedding refuses to cooperate, it may be necessary to sprinkle additional water via a fine rose attached to a watering can or a hosepipe, to wash the mortar/concrete into position around the stones. Where the bedding has stiffened to the point where it will no longer flow, a wet mortar slurry can be prepared and poured between the cobbles, where, it is to be hoped, it will flow and find its own level, sealing in the cobbles in the process.

## LAYING CONCRETE

Concrete is usually regarded as a utilitarian surface – it is not particularly attractive but it is tough, reliable and cheap. There are some techniques used to improve its looks, but many of these need to be undertaken by specialist contractors because concrete is such an unforgiving medium. You are working against time – the concrete is getting harder and less workable with each passing minute and, if the desired effect is not achieved within a relatively narrow period, it never will be. The ability to lay a basic concrete slab is a useful skill for many pavement projects. Concrete slabs are used as paths, patios and driveways, and they are also used as bases for other types of paving. This section looks at how a simple slab is prepared, poured and finished, and also considers a couple of elementary techniques that may be used to impart a slightly more decorative finish.

### Preparation

On the face of it, there is nothing that is much simpler than creating a concrete slab: a flattish base, something at the edges to prevent concrete from flowing away, and some means of levelling it out. How hard can it be? First, concrete is heavy and working it is strenuous and you are working against the clock. And as with so many tasks, the key to success is preparation.

The flat base is important. If concrete is poured on to unprepared ground and some parts are 30mm thick while others are nearer 300mm, there will be problems resulting from the way it cures. For an even cure and the best results the slab needs to be of a regular thickness, hence the need for a flat base. Further, it needs to be of a minimum thickness or it

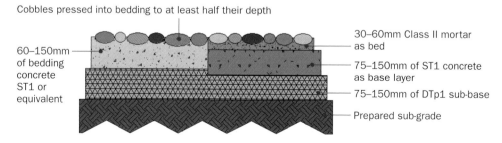

Cobbles pressed into bedding to at least half their depth

60–150mm of bedding concrete ST1 or equivalent

30–60mm Class II mortar as bed

75–150mm of ST1 concrete as base layer

75–150mm of DTp1 sub-base

Prepared sub-grade

*Cross-section for rigid cobble paving.*

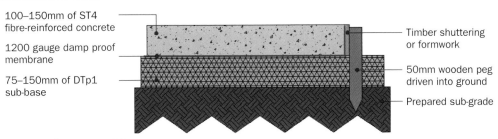

100–150mm of ST4
fibre-reinforced concrete

1200 gauge damp proof
membrane

75–150mm of DTp1
sub-base

Timber shuttering
or formwork

50mm wooden peg
driven into ground

Prepared sub-grade

*Cross-section for concrete slabs.*

will not be strong enough to support its own weight. As a general guide, any footpath or pedestrian-only application will need to be at least 75mm thick, a driveway for typical family cars should be at least 100mm thick, and for heavier uses the slab should be at least 150mm thick.

For pavements or bases concrete slabs are normally laid over a sub-base. This enables the area to be levelled out reasonably accurately and the concrete placed on to a reliable surface. A damp-proof membrane laid over the sub-base prevents the concrete from losing its water content too rapidly, before the cement has been able to hydrate properly. Additionally, a damp-proof membrane protects the underside of the concrete from aggressive ground-water that can accelerate the degradation of a cured and hardened slab.

## Shuttering

The edges need to be retained. This can be achieved by using an edge course of some form and then filling between with concrete. A simple block paver edge course transforms a concrete slab, making it look as though someone has planned it and been determined to create an attractive looking finish. Kerbs, edgings, setts, dwarf walls or almost anything that can be fixed in place may be used as a retainer edge. It should be firm and sound since there is a lot of weight in wet concrete and the last thing that is needed is for a retaining edge to give way.

Temporary edge restraints may also be used. These are put into place to keep the concrete where it is wanted until it has hardened sufficiently to stay put, at which point the temporary restraints, known as formwork or shuttering, can be removed. Timber is a popular choice for shuttering, although preformed shutters are also available. The shuttering needs to be

secure and this is typically done by driving stakes into the ground (or sub-base) at 450–600mm centres.

The top of any shuttering or edge restraint ought to be levelled flush with the required slab, allowing for any necessary falls, which for concrete slabs are usually around 1:80–1:40. Accurate establishment allows edge restraints to be used in the same way as screed rails when it comes to levelling out the concrete, ensuring that all high spots and low spots may be identified and corrected before the concrete sets.

## Suitable Concretes

The type of concrete that is used is obviously relevant. Small batches of less than around 1m³ can be mixed on site, but larger quantities are much better delivered to site as 'Ready Mix'. This guarantees a consistent mix, of uniform strength and slump (and colour) that has to meet certain standards, not least of which is a predetermined strength: no guesswork; no worrying about running out of materials; no trying to remember whether you have put four shovelfuls of cement in the mix or was it five? Further, the delivery wagons can often discharge the concrete direct to where it is required and take away any excess, or bring extra if there is a shortfall.

For more inaccessible sites, such as around the back of a house or at the end of a long driveway, there are concrete pumps that can take Ready Mix from a delivery wagon and disgorge it wherever it is needed. There are also 'mix and move' contractors that bring all the aggregates and cement to site in specially developed wagons which are used to mix the concrete to the required strength and slump. The contractors barrow the concrete to where it is needed and they will even level it out and float finish, if you ask.

A typical concrete suitable for path, patio or driveway use is often referred to as ST4 (Standard

*Tamping a freshly-poured concrete bay using a straight-edge timber as a tamp.*

Mix Design 4). This is roughly equivalent to the old 1:2:4 mix design and has a 28-day compressive strength of around 20N/sq.mm. It is a good, all-round, general purpose concrete, and strong enough to take the sorts of load that may be expected on residential pavements, but not so strong that it is liable to go off in next to no time. A good slump for slab work is 50–60mm, which is reasonably wet and workable, but not too sloppy nor soup-like.

For the construction of slabs, a fibre-reinforced concrete is now preferred to the steel-mesh reinforcement popular in the past. It is considerably cheaper, it requires no special preparation, it is safer (because no steel has to be cut or moved into position) and it is much better at controlling small surface cracks. The small polyester fibres are added to the concrete at the batch plant so that they are thoroughly dispersed throughout the mix. The cost is minimal – at around £5 to £10 extra per cubic metre, which at 100mm depth is less than £1 per square metre, compared with the cost of steel, which for a single sheet of 6mm mesh is around two or three times that price.

## Working the Concrete

Once concrete has been poured into place, it has to be levelled, tamped and finished. Levelling is done with shovels and rakes, spreading the wet concrete and using its flowing nature to persuade it to go where it is required. However, unless a high slump concrete is used, it is unlikely that shovels and rakes will be able to level a slab accurately, thus a screed technique that uses the edges of the slab area as level guides is required.

The screed board is often a long, straight plank, with someone at each end to 'saw' the concrete surface, moving the plank forward a little at a time and redistributing the concrete as necessary. Professional contractors may use a vibrating beam screed, which is essentially a double screed board mounted by a small engine that shakes and vibrates the structure, encouraging the concrete to flow in front of the beam screed as it is dragged across the surface.

The beam screed has the advantage of vibrating the concrete as it screeds it to level, and so any air bubbles trapped within are driven to the surface. When a plank is used, the concrete must be tamped, which means banging the surface with the plank to create vibration within the concrete. The plank must strike the surface along its full length simultaneously otherwise the effect is seriously diminished. When using a small screed board, a single person can control the tamping process but longer plank-like screeds must be controlled by two operatives who will need to operate in synchronous rhythm to ensure that the tamping edge of the plank remains parallel to the concrete surface.

## Finishing

Once tamping or beam screeding is complete, the concrete can be left to set. The surface will be slightly rough and may be marked with small ridges indicating the passage of the tamp or the beam screed. If the concrete is intended as a rough-and-ready surface, or as a base for further paving, this surface should be adequate. However, for a smoother surface the concrete has to be 'finished'. This usually involves polishing the surface by using a float. Commercial projects may use a 'power float', which is a set of float blades mounted on a frame and driven by an engine. These machines produce wonderfully smooth surfaces, but their use is not for the faint-hearted and is probably best left to experts.

The timing of the float work is important; if it is attempted too soon, while the concrete is too wet, the float will leave a series of marks and trails in the surface; if left too late, it may be impossible, or nearly so, to smooth the surface adequately and bring the 'fat' (the fine matrix of sand and cement) to the surface. The correct time to float is when the concrete has firmed up so that tapping the surface does not create any movement or rippling, but polishing with the float brings up the fat while driving down the coarse aggregate. The float is swung around in a series of arcs, to polish and smooth, gradually using lighter and lighter strokes until no marks are left in the surface. The edges of the concrete slab can be 'arrised', which involves forming a rounded profile rather than a square corner that is likely to crumble or break over time.

## Decorative Finishes

One popular finish is the 'brush and trowel' effect. After floating is complete, the surface is lightly marked by drawing a stiff brush across the still plastic concrete to leave a series of fine ridges that will improve traction. An arris trowel (a float with one edge shaped to form an arris) is used to polish the edges of the slab as a decorative detail. Imprint rollers can be run across the floated concrete to impress a series of small depressions, again with the aim of improving traction and creating a decorative finish. Another popular finish is the 'exposed aggregate' look, which involves lightly covering the surface of freshly floated concrete with decorative gravel and using a tamp or a screed board to press it carefully into the surface. As the concrete cures, it will hold most of the decorative gravel fast and create an attractive looking surface. The surplus gravel can be swept off once the surface has hardened sufficiently, usually four or five days later.

## GRAVEL SURFACING

Gravel is regarded as a cheap and simple surfacing, particularly for larger drives on older properties. Success with it depends on understanding of just how it works. A common misconception is that gravel needs to be 50mm or more in depth. This actually produces a 'gravel trap', rather than a gravel drive, as the depth makes it impossible to use without sinking in. Gravel works best when it is used as a surface dressing to something more substantial.

Usually, this is a sub-base, constructed exactly as described earlier and then topped-off with a 25–40mm layer of selected gravel. The gravel will be scuffed off the surface, exposing the sub-base beneath, but this diminishes over time as gravel is driven into the sub-base and so 'blurs' what was the sub-base surface. Choosing a sub-base material similar in colour to the gravel dressing will help, since it makes the scuffing of the early days much less noticeable.

### Size Matters

Choosing the right size of gravel is important: too small and it scatters everywhere, attract cats from miles away, and gets carried into the house on the soles of shoes; too large and it is difficult to walk across, like a pebble beach. Generally speaking, 6–10mm gravel is ideal for foot traffic areas, and 10–18mm for vehicular areas. Angular gravels (those produced by quarrying rock) give a higher degree of

*Concrete is finished by smoothing the surface with a float.*

*Once floated, the surface may be scored lightly using a brush to create tiny ridges running at a right angle to the main direction of travel …*

*… and the an arris trowel can be used to re-smooth the edge and give a bull-nosed profile.*

*Gravel is cheap and simple, but may take a lot of maintenance to keep it where it is supposed to be.*

*Resin-bonding of a decorative aggregate is a relatively simple task suitable for DIY. (DecorDrive)*

interlock and therefore a more stable surface, while rounded gravels remain looser for longer.

## Other Gravels

This 'looseness' of gravels is a common cause of concern. Some people advocate the adding of sand to help 'tighten it up', and, while this does work to some extent, it takes a significant quantity of the finer material to have any real effect. A better solution is to use 'self-binding gravel'. These are usually crushed, angular products with a significant quantity of fines: 10mm to dust would be a typical size, and so they work exactly as does a sub-base, albeit on a smaller scale. Many of the more popular self-binding gravels consist of limestone, such as the legendary Breedon gravel, although there are granites and gritstone versions.

Hoggin, a mix of gravel, clays and sands, is popular in the south of England but virtually non-existent in the north of the country. Again, it should

be laid as a thin surface dressing layer over an adequate sub-base, since on its own it is not really capable of supporting anything much heavier than a wheelbarrow.

## Resin-fixed Gravels and Aggregates

A more modern approach to the problem of loose gravels is to 'glue' them into place. For years, this was done by using bitumen or tar applied to a substrate such as a bitmac surface or concrete slab. The selected gravel was spread over the bitumen/tar and pressed into it with a roller. Over the past decade, the use of polyurethane resins has become more widespread since they adhere better to both the substrate and the gravel and thereby extend the service life of the surfacing. Some of the modern resin-bonding systems have been specifically designed to be suitable for DIY use, and offer a simple yet effective method of creating a natural looking surface with none of the drawbacks normally associated with loose surfacing.

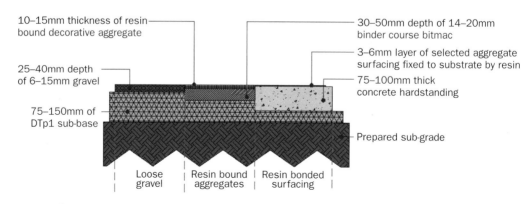

*Cross-section for various gravel-based surfaces.*

174

# CHAPTER 13

# Steps, Ramps, Stepping Stones and Crazy Paving

## STEPS

So far we have looked at contiguous pavements constructed at one level, which is fine for driveways but terraced patios are constructed at a variety of levels that need to be linked by steps. Steps are also popular features at doorways, whether they are necessary or not, as they are perceived to add status to an entrance – they certainly attract the eye when used on large paving projects, as our brains are attuned to scan for anything different, which includes a sudden and pronounced change in level. When it comes to their construction, as discussed in Chapter 3, preplanning is essential: a basic plan should be in hand, so that the work can proceed with the minimum of guesswork.

### Construction

There are dozens, possibly hundreds, of different ways to construct steps using many types of material. As long as the construction is sound, safe and meets the requirements outlined below, it should be perfectly satisfactory. This section considers the construction of the more popular forms, including a single doorway step and a multiple flight, constructed by using readily available paving materials.

Steps constructed from concrete paving materials, from stone such as flagstone, setts or cubes, or from clay pavers and masonry, need to be built upon a concrete base for stability, and each of the units needs to be firmly and securely keyed into that base. For constructed steps, an ST1 concrete is normally adequate as a base, but for cast-*in-situ* concrete steps, a stronger, tougher ST4 mix is more suitable.

Construction always starts at the bottom of a flight and proceeds upwards. The bottom landing may be constructed first, but this is not always essential, and it may be that the lowest riser is the first component to be built. Risers should be vertically plumb and checked with a spirit level. For many steps, the riser will be backfilled with concrete, brought up to a level that will support the tread; but for brick-built steps, this is not always the case, as can be seen from the diagrams on pages 54 and 55.

The tread comes next. This needs to be securely affixed to the riser beneath, if an overlap construction is being used, or keyed-in behind the riser for composite construction. The tread is the component

*Steps should be regular and even. (Stonemarket)*

most directly responsible for supporting the users, so stability and a firm fixing are absolutely essential. The tread should have some endfall to ensure that water cannot collect on its surface, and this too should be checked with a spirit level.

The next riser will normally be keyed-in behind that tread, to prevent it from slipping downwards, but on a 'flag-and-brick' type of construction, the risers are sometimes built on top of the preceding tread component, and so a good mortar with a strong bond strength is vital. Construction continues, with treads and risers being built alternately until the final tread or landing, at the top of the flight, is completed. This may be a tactile paving of some form, even if the steps are not required to be Document M compliant.

Where sidewalls are required, those constructed from masonry are normally built before the construction of the steps, while those formed by using a simple flag on edge are constructed as part of the total step construction, progressing uphill as the steps do.

## RAMPS

As with steps, ramps require preplanning and construction normally starts at the base of the ramp, proceeding uphill in much the same way as with step construction. The ramp needs to have a steady and even gradient, so a taut string line should be used to guide the laying level.

Ramps constructed by using block pavers should have their edge courses laid first, using the taut string lines to guide level, and then have the body paving laid in the customary way. Flagged ramps also benefit from the inclusion of an edge course of some form since this shifts the position of any cuts away from the edges or against sidewalls.

Laying concrete to a ramp is a little more tricky because if the mix is wet enough to work it is probably also wet enough to want to flow back down the gradient, so getting the slump just right is critical. Tamping will encourage the plastic concrete to head downhill, so it is best to work uphill, keeping tamping or vibrating to a minimum.

## STEPPING STONES

These are normally encountered only in a garden setting where occasional access is required and any form of hard landscaping needs to be kept to a minimum. In many cases, stepping stones consist of nothing more than a single flag placed down on bare earth. While this may well suffice, a more stable

*For larger flights, make sure that the treads have ample width. (Tobermore)*

*When it comes to doorsteps, there is no limit to what is possible. (Tobermore)*

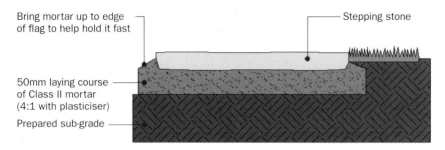

Bring mortar up to edge of flag to help hold it fast

Stepping stone

50mm laying course of Class II mortar (4:1 with plasticiser)

Prepared sub-grade

*Stepping stones need to be held securely in place by bringing the mortar bed part way up the edges.*

and permanent construction may be preferred in situations such as lawns, where the stepping stones may well be trafficked more regularly or traversed by heavier mowing machinery.

The best method to ensure stability is to bed the chosen paving on to a base of fresh concrete. Even if block pavers are being used, a concrete bed is simpler and more effective than an attempt to construct edge courses and a separate sand bed. The concrete should be 50 to 100mm deep and extend beyond the 'stone' by 50 to 100mm, with some of the concrete brought up the sides of the stones as a token haunch that will prevent any lateral slippage of the stone.

The stones should be laid with a touch of fall, if at all possible, to prevent water from accumulating or ponding on the surface. 'Stones' consisting of more than one unit can be cambered so that water is shed to either side or all around, depending on the format. The spacing of stepping stones is never easy. What suits the pace of an adult female is probably something of a stretch for a child and rather too short for an adult male, thus some compromise will be required, unless the stones are of sufficient size that they can accommodate all gaits and paces. The position of each stone can be set out by asking a member of the household to walk the proposed alignment and mark each footfall, which will be the centre of each stepping stone. Alternatively, having the stones at 450 to 750mm centres is a typical compromise – whether it is a compromise because it suits everyone or because it suits no one is debatable.

*Stepping stones need to be carefully spaced. (Stonemarket)*

# CRAZY PAVING

The term 'crazy paving' comes from the crazed appearance of the surface on completion, but it could equally apply to those that undertake a project involving it ... Crazy paving is mistakenly viewed as a simple option for gardens and patios, even for driveways, but it is actually much harder to do, and do well, than it is to lay 'normal' flags, so much so that one notorious contractor somewhere in north-west England would lay full flags and then break them once they were in position, spreading the joints as required and brushing in a slurry mortar. The result looked like exactly what it was: full flags broken up, with the square or rectangular outline of the original, unbroken flag still apparent. This anecdote is included only to emphasize the point that a good crazy paved patio or driveway involves a lot of extra work and is a request dreaded by many in the trade.

The flag pieces are usually delivered to site in a broken and battered state. Builders' merchants sometimes give away broken, plain, concrete flags, although they usually expect money for the broken decorative patio flags since these can usually be returned to the manufacturer for a refund or replacement. Many paving contractors are only too happy to deliver a couple of tonnes of old or broken concrete flags from a nearby job, as long as you do not ask them to get involved in the laying. Broken stone flags are usually sold off as 'random rubble paving' and can attract quite a hefty price.

All crazy paving, whether concrete flags, patio flags or stone, is best laid on a bed of lean-mix concrete (ST1) or 8:1 grit sand with cement. If the pieces are of a regular thickness, a screeded bed may be used, but rubble stone paving may require that individual beds are prepared. The real secret of success with crazy paving is the careful selection of the pieces. It is like a jigsaw puzzle: the best-fitting pieces need to be plucked from the pile and put into place so that joint width is kept to a minimum. Crazy paving with excessively wide joints never looks right: the best results are found on those projects where the joint width is kept to a minimum, with just enough of a gap between pieces to emphasize the non-rectangular, 'crazy' layout. Try not to use too many pieces of

*The key to success with crazy paving is fitting together the individual pieces to the best effect.*

100mm or smaller as gap fillers, but look for larger pieces that can be trimmed to fit.

As there are so many joints, and because they run in so many different directions, it is essential that falls and levels are maintained to drain surface water properly from the pavement, otherwise small puddles will collect here, there and everywhere. Use a straight-edge to check levels continually as the paving is being laid and be prepared to adjust, readjust and re-readjust individual pieces of paving to get the levels just right. Levelling-up non-rectangular pieces is more challenging than the same task using full flags. Pointing is best done by hand, although slurry jointing is much easier. Tooling the joints is difficult because of their non-parallel nature, so the mortar usually has to be smoothed and tidied by using a pointing trowel, working around the angles and curves of each piece of paving.

Good effects can sometimes be achieved by incorporating different colours and textures within a crazy paved surface. The occasional brick, piece of coloured stone, or even items such as seashells and toughened safety glass, can add interest to a surface by emphasizing the 'crazy' design.

# CHAPTER 14

# Completion, Remedial Work and Disputes

## SNAGGING LISTS

It is an unwritten rule of the paving game that no job is ever completely problem-free. Whether you have employed the best contractor in the area or relied on your DIY skills, it is pretty certain that, come the end of the main construction work, there will be a list of minor faults, issues, annoyances and items that have been awaiting completion as soon as you or the contractor get around to it. This is known as the 'snagging list'. It is a worthwhile exercise compiling such a list, even if it has been a DIY effort or just for your own use, to remind you to remind the contractor. Trying to remember all the little fiddly things that require attention is virtually impossible, but if they are jotted down, the schedule can be planned, the materials arranged and each task ticked off as it is completed. For DIY-ers, a snagging list is simply an aide-memoire; when a contractor is involved, touring the work and drawing up the list together can help both parties to recognize and agree just what remains outstanding.

Most competent contractors will not take umbrage at the suggestion of a snagging list. The better contractors will have a mental list of what they need to look at and rectify before submitting the final invoice, and so formalizing it should be no problem. To compile a mutually agreed list during a pre-arranged inspection of the site gives the client an opportunity to point out anything that has been irking him or her, and allows the contractor to see exactly what is meant, as well as the chance to explain just how insignificant it is and how easily it can be corrected. An agreed list is much more likely to yield results than some ill-considered ultimatum of demands from a client not fully familiar with construction work and the language of the trade.

### Checking the Work

So what should be checked? What is a legitimate concern and what is nit-picking? Something that annoys you and your partner intensely may well have been completely overlooked by the contractor, while that section of haunching at the back of a kerb line that has been missed – which you did not even know was missing – may well be the number one item on the contractor's agenda.

The type of paving has a major bearing on what should be checked. A block-paved patio may well have a few empty joints, but, as was explained earlier, with block-paving construction this is only to be expected. However, empty joints on a Yorkstone flagged patio are a genuine cause for concern.

Consider first the overall appearance of the work: is it reasonably neat and tidy? Can you discern how the falls and gradients will work to drain the surface? Are there any obvious low spots or humps? Are there any broken, damaged or plain ugly paving units that need to be replaced? How has the paving been tied in to the public footpath or the highway at the threshold? Has all the rubble and debris been cleared? Has any damage to grass verges been made good? Are you pleased and satisfied? Next, look at the standard of workmanship: are the joints that should be straight actually straight, or are they askew? Are the cuts straight and tight, or are there gaps? Do the flags rock when walked upon? Does water 'pump' up from joints in block paving? Are there any mortar or concrete stains? Are the edges firmly secured? Is the finishing around gullies and access covers acceptable?

Disputes between contractor and client often arise over 'subjective' matters. The client thinks one section of paving is stained while the contractor insists that it is a temporary phenomenon, or there is a difference of opinion regarding whether a step outside the kitchen door looks too small. Within reason, the contractor should always try to provide what the client wants, even if it goes against the contractor's better judgement. They should certainly consider the client's opinion and, if there is a valid structural reason why a particular request is impracticable, then it should be put in writing, if necessary; but it is the client's property, and if they want a step twice the width of the doorway that will actually prevent the back gate from being fully opened, and make it almost impossible to get the wheelie bin out, then they are entitled to have it. By putting in writing the reasons for not accepting a particular request, the contractor is effectively denying any responsibility for the work if and when it is found to be wrong.

However, there are some 'issues' that can be shown to be right or wrong. Falls can be examined with a spirit level and documented by using an automatic level; high spots or hollows can be proven by use of a straightedge – any dip or high spot of more than 6mm is significant and requires remedial action; alignment or straightness can be verified by use of taut string lines.

In general, modular paving such as flags, block pavers and setts can be repaired without too much trouble. Individual units can be lifted out and put back or replaced as necessary. However, any repairs to monolithic surfaces, such as concrete, tarmac, pattern imprinted concrete (PIC) and resin systems, will involve permanent and possibly obvious patching. This is why many tarmac and PIC contractors are so reluctant to carry out repairs – it is quite likely that a repair will result in the paving looking even worse, especially if the problem area is in a central or prominent location within the pavement.

To return to the snagging list: if a catalogue of 'problem items' can be agreed, it would be reasonable to expect the contractor to rectify these before final payment and for such remedial work to be undertaken before they leave the site. Promises to return 'in

*Poor cutting-in: joints have been left too wide and filled with sand, small 'darts' have been used in place of in-board cuts and the mortar pointing to the kerbs is messy.*

*No matter how you try, patches in tarmac always look like patches in tarmac!*

a couple of weeks' are worthless, and a small retention (money held back until the satisfactory completion of the work) is unlikely to provide sufficient leverage to persuade a truculent contractor to come back and fix those two broken flags. So, the best strategy is to resolve all the items on the snagging list before settling-up.

## Retentions

To impose a retention sum without prior agreement with the contractor is unfair, although some contractors will agree to a short 'defects period' of, say, three months. It is much better to agree a retention sum before work begins so that both parties know what to expect. On many commercial projects, a retention sum of 5 per cent is standard practice. This is reduced to 2.5 per cent on 'practical completion', which is the point when the contractor and the client agree that the work is, to all intents and purposes, finished. The retention is paid off after an agreed defects period, usually six to twelve months after practical completion. During this period the contractor is expected to make good any defects that can be shown to be their responsibility.

While 2.5 per cent of a £250,000 commercial project might be a big enough lever to persuade the contractor to carry out the remedial work, the same percentage of the £5,000 cost of a private driveway is unlikely to carry much clout, and, rightly or wrongly, some contractors would be happier to forgo the outstanding £125 rather than send around a repair gang for half a day. Bear this in mind if you are planning to incorporate a retention sum in the contract negotiations.

## Dealing with Disputes

Provided that there is no conflict with the terms of any contract that has been signed, if the client believes the work to be sub-standard or unsatisfactory, and the contractor is uncooperative or truculent, payment may be withheld until the work is brought up to an acceptable standard. The contractor should be notified in writing of the intention to withhold some or all of the sum due, the reasons for non-payment should be documented in as much detail as is possible and there should be a statement to the effect that it is not the intention of the client to avoid payment and that, once the work has been corrected to a satisfactory standard, any money owing will be paid.

If a contractor believes that a client is being deliberately awkward or attempting to get out of paying for work done, they can issue a written demand for payment, stating that, in their opinion, the work is complete and of an acceptable standard of workmanship, and giving the client a time limit in which to comply.

It is unreasonable for a client to withhold more than 50 per cent of the project sum, unless there are strong reasons for so doing. For the vast majority of projects at the 'more-or-less-finished' stage, 50 per cent will cover the cost of materials and possibly also some of the labour costs borne by the contractor. A

sum of 25 to 50 per cent is usually sufficient to persuade a contractor to comply with the remedial requirements. For projects that have relied on 'stage payments', the final instalment is often 25 to 40 per cent, which is a reasonable sum to withhold until satisfactory 'practical completion'.

Sadly, too many disagreements spiral downwards into open warfare. The client refuses to meet the contractor halfway, or wants to withhold the full cost of the works until half a metre of edging kerb has been replaced, or the contractor is adamant that the 'bird bath' of a puddle outside the front door will disappear over the next week or so but is at a loss to explain the physics of self-correcting tarmac. Voices are raised, doors are slammed, threats are issued, blood pressure rises and nothing is sorted out. Verbal arguments rarely resolve a disagreement.

Disputes should always be put in writing. This ensures that the reasons for the dispute are documented, thought through, and explained in terms understood by both parties. It also helps should the dispute end in the local court since it provides documentary evidence of who said what and when, and what was offered and why, and eliminates the he-said-she-said verbal jousting that does no favours to either side. Letters should state the reason or reasons for the dissatisfaction, what remedial action is required to resolve the problem, and what actions will ensue should the matter not be resolved. A reasonable period should be allowed for any remedial work to be completed or for money owing to be paid, seven to fourteen days is usually deemed to be fair.

Regrettably, some projects result in a total breakdown of communication and loss of faith between client and contractor. In such cases, a third party may be required to resolve the dispute. If the contractor is a member of a trade body there may be an arbitration service available; however, the majority of driveway and patio contractors, in common with most other contractors in the home improvements industry, have no trade or professional affiliations.

It may be possible to request a trade body or a professional association to appoint an arbitrator, even if the contractor is not a member, but the fees and costs incurred are likely to be significant. It is difficult to obtain a professional opinion for less than around £400 plus VAT. However, there is nothing to be gained by bringing in an arbitrator unless both parties agree to abide by the findings. In some cases both parties bring in their own 'expert' and agree that the experts should assess the site and then decide between themselves what is to be done to resolve the problem, with both parties abiding by their joint recommendations.

The local trading standards office may be able to suggest an arbitrator or some suitably qualified person to intercede and they should always be informed about cases of sharp practice or 'cowboy' activity so that they can keep an eye on the local renegades. Although trading standards are often regarded as having few, if any, teeth, the local office can and do act if a trader is brought to their attention on more than one occasion, but they do rely on information from the public.

Legal action should always be the last resort as the only guaranteed winners are the lawyers. However, there are cases where the small claims court or the county court is the only route forward. Although justice is usually delivered, it tends to be delivered some time after the fact, and the intervening period can be extremely stressful and lead the client to question just whether it was all worth it. Even if a case goes to court and is adjudged one way or the other, enforcing the judgment is not always straightforward and it may require further visits to the court to persuade it to send in the bailiffs or other enforcement officers.

In summary, if the client and the contractor can find some common ground between themselves and rely on fairness, integrity and common sense, most disputes can be resolved without needing anyone to lose face, money or their temper. Although hindsight is a wonderful gift, if a client and a contractor cannot communicate, perhaps they should never have paired up in the first place. The easiest way to avoid a dispute is to use a simple, no fuss, plain English contract before the work starts, then both parties will know what is expected.

## AFTERCARE AND MAINTENANCE

Once a paving project is complete, how should it be looked after? Despite the claims of some of the more dubious paving and surfacing companies, there is no such thing as a 'maintenance-free' driveway or patio. You may decide that your patio or driveway is a maintenance-free area – by neglecting it, but if you prefer to safeguard your investment and ensure that it is kept in tip-top condition for as long as possible, some regular, elementary maintenance will be necessary.

### The First Three Months

This is a 'running-in' period for most paving and surfacing. The jointing sand used with block paving and some flags will settle and should be topped up. The better quality contractors will do this as part of their aftercare regime, but, sadly, only around one in forty contractors ever bothers.

Pointing or other jointing should be checked and any sign of settlement or movement should be noted and earmarked for remedial DIY work or brought to the attention of the contractor. Keep traffic speeds to a minimum on a new tarmac surface since the 'volatiles' used to keep the material fluid and workable when the surface was originally laid can take a couple of months or so to escape. Check gullies and linear channels to ensure that they are functioning properly and not blocked by sand, sediment or other building rubble.

With concrete products, and with some clay pavers, efflorescence may appear during this period. This white, powdery substance is an unavoidable fact of life for many concrete products. The better manufacturers tailor their manufacturing process to minimize its occurrence, but it is impossible to eliminate completely, and virtually every manufacturer has an

*Efflorescence affects all kinds of paving but is temporary.*

efflorescence disclaimer somewhere in their small print. The good news is that, apart from being a minor eyesore, it is completely harmless and will disappear eventually. There are 'efflorescence removers' available from builders' merchants and DIY stores, but they are what they say – removers, not cures. They will probably improve the appearance for a couple of weeks but there is a good chance that the bloom will be back again within the month.

Try not to use power washers on the paving during its early life. They should be shunned completely for block paving and other sand-jointed paving until the joints have 'self-sealed', which normally takes three months as a minimum. Keep any cleaning procedures simple: sweeping with a stiff brush to remove any detritus and litter, and swilling down with a bucket of soapy water, if it is absolutely necessary.

## Three-to-Six Months

The 'running-in' should be complete by now. The jointing sand should have stabilized and a layer of very fine material will probably have formed a seal to the surface of the joints. The contractor or DIY-er should have rectified any minor settlement, and it is likely that the first of the wind-distributed weeds will be trying to set up home somewhere on the surface. Regular sweeping and the removal of any detritus that accumulates can usually control this, but it may be worth considering the use of a sealant.

### Sealants

Sealants should be applied to concrete block paving, clay pavers or concrete flags only after any efflorescence has vanished and after the jointing is known to be firm and permanent. This is normally not less than three to six months after the original

*Sealants not only protect the paving, they reduce the opportunity for weeds to grow and enhance the colours. (Resiblock)*

construction. Applying a sealant too soon after construction can 'seal in' the problems and do more harm than good.

Sealants can protect a surface from accidental staining, but they can also eliminate the growth of algae as well as binding together any loose jointing sand. There are a number of general purpose paving sealants available, but these tend to rely on being a 'lowest common denominator' product that is capable of sealing concrete, stone, slate, terracotta, ceramics and anything else, rather than being a specialist product developed to suit the individual characteristics of the various materials that are used for paving. The best results are usually obtained with material-specific sealants, rather than the jack-of-all-trades offerings: sandstone sealants for sandstone, concrete sealants for concrete, and resin seal-coats for resin.

Concrete block-paving sealants, for example, come in a number of different guises, but the two most popular formats are acrylic-based and polyurethane-based. Acrylics tend to be cheaper and may need to be re-applied every two or three years, while urethanes are tougher and should give a minimum of four to six years of service before needing attention.

There is some difference in the way the various sealants protect paving. Cheaper products usually work by depositing a protective film on to the surface of the paving, which then hardens over a few hours. Such sealants are prone to abrasion – they become worn, especially on projects where cars and people are constantly using the same routes – front door to garage, and garage to driveway threshold. The better quality sealants usually penetrate the paving to a depth of a few millimetres and so offer a thicker protective barrier that cannot be abraded. Cheaper sealants also have a tendency to yellow with age. This is a result of exposure to UV, which degrades the material over time – so look for 'UV-stable' or 'non-yellowing' formulations.

The effect a sealant may have on the appearance of the paving should also be considered, and it is always best to test a product on a discreet corner or some spare paving before daubing it all over the driveway. Some sealants impart a very definite, glossy, wet look that may not be to everyone's taste. They do, however, enhance the colouring, which is important with some surfaces, such as PIC. There are also low-sheen (sometimes referred to as 'matt finish') sealants that give some colour enhancement without looking too glossy. Other sealants have little or no perceptible effect on the surface's appearance.

For stone flags there are different sealants for different types of stone, based on its porosity, so make sure that the correct product for your stone is used. Some stone sealants are liquids to be painted on, while others are wax-like and need to be rubbed in, rather than applied by brush. While a colour-enhancing wet look or satin sheen may be perfectly acceptable for stone or concrete flags used as the internal flooring for a kitchen or conservatory, a more subdued finish may be preferable on a patio or driveway, so choose carefully. Also be aware that some stone-sealing products, especially the waxes, are not always suitable for outdoor use. For liquid sealants, application is usually simple: ensure that the paving is clean and then use a roller or a squeegee to apply the liquid evenly over the entire surface. Many sealants require two coats, with a time delay between applications to allow the first or 'primer' coat to cure properly. Coverage varies significantly, depending on the porosity of the material, but figures of 3 to 6m² per litre per coat are typical. The manufacturers should be able to provide more detail.

Bear in mind that sealants are irreversible. Once applied they cannot be removed (there are one or two exceptions to this, but removal can cost between four and ten times the cost of application). Further, some types of sealant may be incompatible: some of the polyurethane-based liquid sealants do not properly penetrate or adhere to surfaces previously treated with a wax-based product. When re-sealing a previously sealed pavement, check with the manufacturer before parting with your money.

## Twelve Months and Beyond

By this time maintenance should be reduced to little more than a monthly brushing and the quarterly application of a weedkiller, if the site is plagued with weeds or in a damp or shady spot that is prone to mosses, algae and lichens.

Power washers may be used, but the jet of water must be kept at a shallow angle to the paving to minimize any risk of damage. It must never, ever be

*It is important to keep the lance at a shallow angle and not aim at the jointing when cleaning pavements with a power washer.*

directed on to sand jointing as it will simply blast out the sand, rendering the pavement vulnerable to movement and settlement. Any sand accidentally removed *must* be replaced as soon as the surface is dry. Similarly for mortar jointing – any accidental removal must be made good, otherwise the structural integrity of the paving will be compromised.

Gravel surfaces may be more difficult to maintain because of their inherently loose nature. Leaf litter and other detritus are best removed by using a spring-tined lawn rake; weeds that have created a home for themselves can be evicted with a hoe and a weedkiller applied to discourage their return. Plain or textured concrete can be cleaned with a stiff brush and soapy water, although a power washer should do no harm, if that is preferred. Tarmac and PIC need more care, since a power washer may damage the surface or dislodge loose material, particularly as the paving ages and deteriorates. PIC benefits from regular resealing since this helps to maintain the depth of colour and protect the vulnerable topcoat. Sadly, many PIC contractors last only a couple of years or so in the trade and therefore it may be necessary to undertake this vital work as a DIY task. Discoloured, stained or faded, tired-looking tarmac can be given a new lease of life by applying a resurfacing compound. Some of these products are little more than glorified paints, but the better ones chemically bond to the original bitumen content of the surfacing, strengthening it and bringing a new depth of colour to make it look like new.

# Useful Links

## www.pavingexpert.com

Much more information on all aspects of paving, including many types and materials not covered in this book may be found on the author's website, www.pavingexpert.com. There is a lively and friendly discussion forum featuring many contributions from the author along with those of several experienced professional paving contractors and keen DIY-ers, all answering queries and offering advice to everyone from a DIY novice to fellow contractors dealing with a new problem. Several paving manufacturers and suppliers also encourage their staff to contribute helpful tips and pointers to further information.

## MANUFACTURERS

Acheson-Glover (block pavers and flags)
Fivemiletown, Co. Tyrone, Northern Ireland
tel: 028 8952 1275; www.acheson-glover.co.uk

Aco Technologies (linear drains and ground improvement products)
Shefford, Bedfordshire
tel: 01462 810223; www.aco-technologies.com

Baggeridge Brick plc (clay pavers and accessories)
Dudley, West Midlands
tel: 01902 880555; www.baggeridge.co.uk

Blockleys Brick Ltd (clay pavers and accessories)
Telford, Shropshire
tel: 01952 251933; www.blockleys.com

Border Stone (decorative aggregates)
Welshpool, Powys
tel: 01938 570375

Bradstone (decorative garden paving)
Ashbourne, Derbyshire
tel: 01335 372289; www.bradstone.com

Brett Landscaping (block paving, patio flags, decorative aggregates)
Cliffe, Kent
tel: 01634 221801; www.brett.co.uk

Charcon (block paving, flags and kerbs)
Ashbourne, Derbyshire
tel: 01335 372222; www.charcon.com

ClarkDrain (manhole and access covers, recess trays and linear drains)
Yaxley, Peterborough
tel: 01733 765315; www.clark-drain.com

Creative Impressions (PIC tools, texture mats, colours and sealants)
Bamber Bridge, Preston
tel: 01772 335435; www.creative-impressions.com

DecorDrive (resin-bonded surfacing for driveways)
Harlow, Essex
tel: 0800 731 2382; www.decordrive.co.uk

Ennstone Johnston plc (Breedon gravel)
Breedon on the Hill, Derbyshire
tel: 01332 862254; www.ennstone.co.uk

Formpave Ltd (block pavers, Pennant stone)
Coleford, Gloucestershire
tel: 01594 810577

Gardner-Gibson Inc. (tarmac resealing products)
Tampa, Florida, USA
tel: +1 800 237 1155; www.gardner-gibson.com

Hepworth Drainage (concrete, clay and uPVC drainage manufacturers)
Sheffield, South Yorkshire
tel: 01226 763561; www.hepworthdrainage.co.uk

Johnsons Wellfield Quarries Ltd (creamy Yorkstone flags and walling)
Huddersfield, West Yorkshire
tel: 01484 652311; www.johnsons-wellfield.co.uk

Longborough Concrete (patio flags and edgings)
Moreton-in-Marsh, Gloucestershire
tel: 01451 830140; www.lonstone.co.uk

McMonagle Stone (Donegal quartzite and sandstone)
Co. Donegal, Ireland
tel: +353 (0) 74 973 5061; www.mcmonaglestone.ie

Marshalls Ltd (flags, blocks, kerbs, walling and natural stone)
Halifax, West Yorkshire.
tel: 01422 366666; www.marshalls.co.uk

Netlon Tensar Ltd
Blackburn, Lancashire
tel: 01254 262431; www.tensar.co.uk

Resiblock Ltd (sealants for block paving)
Basildon, Essex
tel: 01268 273344; www.resiblock.com

Riverside Reclamation (reclaimed stone paving)
Bolton, Lancashire
tel: 01204 533141; www.stoneflags.co.uk

Roadstone Ltd (blocks, kerbs and flags)
Dublin 24, Ireland
tel: +353 (0)1 404 1200; www.roadstone.ie

Rock Unique Ltd (stone for landscaping)
Sevenoaks, Kent
tel: 01959 565608; www.rock-unique.co.uk

Stone & Style
Zutendaal, Belgium B-3690
tel: +32 89 61 00 33; www.stone-style.co.uk

Stone Flair (patio paving and natural stone)
Newark, Nottinghamshire
tel: 0870 600 9111; www.stoneflair.com

Stonemarket Ltd (patio flags and walling)
Ryton-on-Dunsmore, Warwickshire
tel: 024 7651 8700; www.stonemarket.co.uk

Stonescape (reclaimed and new stone setts, cubes and flag paving)
Wigan, Lancashire
tel: 01942 866666

SureSet UK Ltd (resin supplies and contractors)
Warminster, Wiltshire
tel: 01985 841180; www.sureset.co.uk

TDP Ltd (geotextiles)
Wirksworth, Derbyshire
tel: 01629 820011; www.tdpltd.com

Tobermore Concrete Products Ltd (blocks, flags, kerbs, natural stone and walling)
Tobermore, Co. Londonderry, Northern Ireland
tel: 028 796 42411; www.tobermore.co.uk

Wavin Plastics Ltd (uPVC drainage systems)
Chippenham, Wiltshire
tel: 01249 766 600; www.wavin.co.uk

Westminster Stone Company Ltd (concrete and stone patio paving)
Ellesmere, Shropshire
tel: 01978 710685; www.westminsterstone.com

# TRADE ASSOCIATIONS

BALI (The British Association of Landscape Industries)
Stoneleigh Park, Warwickshire
tel: 02476 690333; www.bali.co.uk

Concrete Society (promoters of all things concrete)
Crowthorne, Berkshire
tel: 01344 466007; www.concrete.org.uk

Construction Industry Publications
Sheldon, Birmingham
tel: 0121 722 8200; www.cip-books.com

Interlay (The Association of Block Paving Contractors)
Leicester
tel: 0116 222 9840; www.interlay.org.uk

Interpave (Concrete Paving Producers Association)
Leicester
tel: 0116 253 6161; www.paving.org.uk

Irish Concrete Society (Irish body for all things concrete)
Drogheda, Ireland
tel: 00 353 (0)41 9876466; www.concrete.ie

The Landscape Institute (the Chartered Institute for
British landscape architects)
London
tel: 0207 299 4500; www.l-i.org.uk

Water Research Centre (research centre for water
management and drainage)
Swindon, Wiltshire
tel: 01793 865000; www.wrcplc.co.uk

## CONTRIBUTING CONTRACTORS

John Higgins
Aintree Paving, Liverpool
tel: 07989 376123

Ken Culshaw
KGC Paving, Leigh, Lancashire
tel: 07932 673271

Original Stone Paving Company
Wrexham, Clwyd
tel: 01978 661000;
www.the-original-stone-paving-company.co.uk

## FURTHER READING

*Building Regulations, Part H – Drainage*

*Building Regulations, Part M – Access and Facilities for
Disabled People*

*British Standard BS8301 – Code of Practice for Building
Drainage*

*British Standard BS6367 – Code of Practice for Drainage of
Roofs and Paved Areas*

*Sewers for Adoption (5th edition)*

*British Standard BS 7533-2:2001 Pavements constructed
with clay, natural stone or concrete pavers* Guide for the
structural design of lightly trafficked pavements.

*British Standard BS 7533-3:1997 Pavements constructed
with clay, natural stone or concrete pavers* Code of practice
for laying precast concrete paving blocks and clay pavers for
flexible pavements.

*British Standard BS 7533-4:1998 Pavements constructed
with clay, natural stone or concrete pavers* Code of practice
for the construction of pavements of precast concrete flags
or natural stone slabs.

*British Standard BS 7533-6:1999 Pavements constructed
with clay, natural stone or concrete pavers* Code of practice
for laying natural stone, precast concrete and clay kerb
units.

*British Standard BS 7533-7:2002 Pavements constructed
with clay, natural stone or concrete pavers* Code of practice
for the construction of pavements of natural stone setts and
cobbles.

*British Standard BS 7533-8 Guide for the structural design of
lightly trafficked pavements of precast concrete flags and
natural stone slabs*

*British Standard BS 7533-10 Pavements constructed with
clay, natural stone or concrete pavers. Guide for the structural
design of trafficked pavements constructed of natural stone setts*

*British Standard BS-EN206-1 Concrete: Specifications,
performance, production and conformity*

*Specification for Highway Works*

*Precast Concrete Paving: a design handbook* (Interpave)

# Glossary

**Aggregate** any loose, granular material used in the construction of a *pavement*; sand, gravel, crushed stone, cement and pebbles are all aggregates.

**Arris/arrises** the edges of a block, flag or slab; the 90-degree angle between the top face and the side faces; can be square, *chamfered* or rounded.

**Body** term used to describe that paving lying inside or surrounded by the edge courses.

**Bogens** arc or arches of setts, popular as a laying pattern in Europe.

**Bush-hammering** a *secondary process* that is used to impart a texture to a paving unit; a multi-point hammer, not too dissimilar to a steak-tenderizer, is used to 'distress' the surface and create a distinctive 'worn' appearance.

**Camber** the rounded, arched or convex profile given to roads and other pavements to assist drainage.

**Chamfered** having the *arris* shaped to a 45-degree profile, so avoiding sharp 90-degree edges that are prone to *spalling*.

**Channelization** the effect where a pair of parallel ruts develops in a pavement due to the repeated trafficking of a driveway that has been poorly constructed; usually seen on narrow drives where the same 'tracks' are used every time cars enter or leave.

**Cobbles** individual rocks, large pebbles or small boulders that have been rounded by the action of rivers, seas or glaciers; unlike *setts*, cobbles are not normally *dressed* but are used as found.

**Cubes** dressed stone *setts* that have identical measurements to all three dimensions.

**Delamination** phenomenon whereby stone or concrete degrades into a series of individual layers, like a deck of cards, with the top layers flaking off due to wear or weathering.

**DPC** damp-proof course, an impermeable layer of material built into walls to prevent damp rising up from the ground; DPCs are typically 150mm higher than external ground or paving levels.

**Dressed** usually refers to prepared stone that has been chiselled or sawn or in some way 'tidied-up' before being used.

**Edge courses** bands of paving materials laid at the edge(s) of a pavement and often used to *restrain* the *body* of the paving.

**Fettled** mason's term meaning *dressed*, fixed or improved in some way; stone flags with unsawn edges are often described as being 'quarry fettled', that is, the edges are cut using chisels or cropping machinery.

**Fines** dust or sand-sized small particles found in a crushed *aggregate*.

**Flame-texturing** a *secondary process* that is used to impart a texture to a paving unit; a high-temperature flame is played over the surface of a stone flag and the rapid rise in temperature causes small-scale, localized *spalling* that creates the distinctive texture; it does not work with all types of stone and cannot normally be used with concrete products.

**Flush** at the same level as an adjacent surface; ideally, paving units will be flush with their neighbours, avoiding any lips or trips that could be dangerous; units that are not flush may be low or *proud*.

**Grinding** a *secondary process* that is used to impart a texture to a paving unit; the surface of the paving is ground down to create an exceptionally smooth but non-slip surface that reveals the inner structure of the (usually decorative) concrete or stone.

**Interlock** pattern or arrangement of blocks that has no continuous joints spanning the entire *pavement*; each block or sub-section of blocks is locked into place by changing the alignment and arrangement of the pattern.

**Longitudinal** lengthways – running up and down the direction of travel, rather than from side-to-side, which is known as *transverse*.

**Modular** units that can be used together, linked or connected in some way to create a larger surface or unit without the need for any cutting; Lego blocks are modular.

**Pavement** any hard-paved surface used for foot or vehicle traffic; in common parlance, the term is often used to mean footway, but in the trade and in civil engineering it applies equally to footways, pathways, driveways, bridleways, highways and patios; all have constructed surfaces and are therefore classed as pavements.

**Pitched stone** stone that has been roughly shaped using chisels to remove any lumps and bumps; also referred to as 'rough *dressed*'.

**Proud** standing too high, sticking-up, forming a lip or a trip on a pavement.

**Restraint** holding firmly in place, usually refers to *edge courses* which are said to 'restrain' the body of the paving.

**Riven** refers to stone that has been 'split away' from the parent rock, leaving a roughened, natural surface that is often mimicked by reproduction flags.

**Secondary processes** texturing or other techniques applied to paving products after the initial manu-facture or cutting; includes *shot-blasting, bush-hammering, flame-texturing* and *grinding*.

**Setts** rectangular blocks of *dressed* stone.; sometimes mistakenly described as cobbles but setts are very definite blocks or *cubes*, rather than rounded stones.

**Shot-blasting** a *secondary process* that is used to impart a texture to a paving unit; creates a roughened texture by peppering the surface with small steel balls that break off small pieces of concrete or stone on impact.

**Slump** a measure of the wetness of a concrete or mortar.

**Spalling** chipping or damage done to the face or edge of a paving unit that often leaves an untidy edge or face and/or reveals the internal structure of the material.

**Sub-base** layer or crushed stone beneath a *pavement*, sometimes referred to as 'hardcore'.

**Sub-grade** the bare earth; the ground after any vegetation and top-soil have been removed.

**Substrate** an underlying surface to which a covering layer is applied.

**Transverse** crossways, running across the direction of travel, from side-to-side, rather than up and down, which is known as *longitudinal*.

**Triangulation** the process of defining the position of any point on a plan by reference to at least two other points.

**Upstand** kerbs 'stick-up' above the pavement, this is upstand or 'check' and is intended to keep traffic in place and prevent it from straying on to other areas.

**Wet-casting** method of manufacturing concrete flags by pouring a wet concrete into a mould.

# Index

3-4-5 Triangle  91

Access chambers (ACs)  122

Ballast, as sub-base  82
Base course  85
Basketweave, setting out  144
Bedding layer
    definition  86
    for flags  154
Block Paving
    concrete, manufacturing method  16
    laying  142
    patterns  19, 143-144
    secondary processing  20
Bogens  166
Buttering  156

Capping layers  130
Cartaway capacities  130
Cell matrix  32
Clay pavers  20
Cobbles
    overview  28
    laying  169
Collimation, line of  112
Compaction to refusal  132, 138, 152
Concrete
    laying  170
    mixes  87
Contractors, choosing  65
Costing breakdown  69
Crusher run, as sub-base  82
Crazy Paving  179
Cubes, construction  166
Cutting flags  161
Cutting-in block paving  148

Damp Proof Course (DPC)  106, 111
Damp Proof Membrane (DPM)  171
Density of materials  129
Dipping  132, 141
Disabled access (Document M)  53
Dished channels  118
Disputes  182
Drainage
    backfilling  124
    connecting to  122
    gradients/falls  48, 122
    planning and setting out  106
    systems (Foul, Surface, etc)  119
Driveways, widths and turning circles  45
Dry mix mortar/concrete  88
DTp1 (Type 1)  82

Edge courses, laying  133
Edgings  42
Efflorescence  183
Estimates, obtaining  66
Excavation  129

Falls  48, 51
Flags (flagstone, slabs)
    cutting  161
    laying individually  153
    layout patterns  160
    manufacturing methods  24
    stone flags  27
Formation level  81, 130

Geotextiles  83, 131
Gratings, grids and gullies  58, 116
Gravel
    laying  173
    types  30

Grouting 163
Gullies 116

Hardcore, as sub-base 83
Haunching 85, 136
Herringbone, setting-out 143
Hoggin 32

In-board cutting 148, 161
Individual bedding 153
Inspection chambers (ICs) 106, 122

Jointing sand 151

Kerbs
    definition 42
    laying 133

Landscape fabrics 83
Laying course
    definition 86
    materials 138, 154
Levels, establishing 105, 112
Levels, types of 105
Linear channel drains 58, 109, 118

Maintenance 183
Manhole covers 57
Moist mix mortar/concrete 86
Mortars
    for buttering and pointing 156
    gun-applied for pointing 165
    mixes 87
    polymeric 165
Mowing Strips, use of 42

Paving layer 88
Permeable/porous paving 116
Perpendicular lines, establishing 91
Pitch for jointing setts and cubes 169
Planings, as sub-base 82
Plate compactor (Wacker plate) 79, 132, 138
Pointing 162
Polymeric jointing materials 165
Ponding 106
Power saws 79, 150, 161
Pythagoras' theorem 91

Quarry waste, as sub-base 82
Quotations, obtaining 66

Ramps 176
Random layout rules 160
Rodding eyes 122

Sand, bedding 86
Saws, cut-off/power/concrete 79
Scalpings, as sub-base 82
Screeding
    preparation 138
    tools 76
Semi-dry mix mortar/concrete 88
Setts 166
Skips 72, 129
Slurry mix mortar/concrete 87
Slurry jointing 164, 168
Snagging 180
Soakaways 125
Spot bedding 154
Stepping Stones 176
Steps
    construction 175
    terminology 52
Stretcher course, setting out 144
Sub-base
    definitions and materials 81
    estimating quantity 71
    laying 130
Sub-grade
    definition 80
    improving 83
SUDS 120, 125
Surveying 59

Tangent points, definition 96
Tarmac surfacing (bitmac) 35
Terraces 51
Triangulation 61

Units – metric and imperial 39

Water/cement ratio 87
Wet mix mortar/concrete 87